Routledge Revivals

Urban Land and Property Markets in Italy

Originally published in 1996, Urban Land and Property Markets describes the intricacies of the Italian urban planning system, and the interconnections between the property sector, the national economy, and recent historical developments, including the new challenges facing Italy after the early 1990s collapse of the party system. The book's underlying thesis is that property values are ultimately created by urban planning and investment in infrastructure. Negotiations between local government and developers focus on three basic issues: the ultimate use of urban land, the quantitative control of development via planning permissions relating to city master plans, and the nature of public investment to support growth and property values.

Urban Land and Property Markets in Italy

by Gastone Ave

Routledge
Taylor & Francis Group

First published in 1996
by UCL Press Limited

This edition first published in 2018 by Routledge
2 Park Square, Milton Park, Abingdon, Oxon, OX14 4RN
and by Routledge
711 Third Avenue, New York, NY 10017

Routledge is an imprint of the Taylor & Francis Group, an informa business

Publisher's Note
The publisher has gone to great lengths to ensure the quality of this reprint but points out that some imperfections in the original copies may be apparent.

Disclaimer
The publisher has made every effort to trace copyright holders and welcomes correspondence from those they have been unable to contact.

A Library of Congress record exists under LCCN: 96225563

ISBN 13: 978-1-138-58908-7 (hbk)
ISBN 13: 978-0-429-49186-3 (ebk)

Urban land and property markets in Italy

Urban land and property markets in Italy

Gastone Ave
University of Ferrara

UCL
PRESS

First published in 1996 by UCL Press.

UCL Press Limited
University College London
Gower Street
London WC1E 6BT

and
1900 Frost Road, Suite 101
Bristol
Pennsylvania 19007-1598

ISBN: 1-85728-053-9 HB

British Library Cataloguing in Publication Data
A catalogue record for this book
is available from the British Library.

Typeset in Times.

CONTENTS

FOREWORD TO SERIES

The idea of publishing this series of books on the different national urban land and property markets of Europe was inspired by a five-country research project on the functioning and framework of urban land and property markets. This project, known as the EuProMa Project, was commissioned by the German Federal Ministry for Regional Planning, Building and Town Planning (Bundesministerium für Raumordnung, Bauwesen und Städtebau, or BMBau), and was undertaken at the Faculty of Spatial Planning (Fakultät Raumplanung) of the University of Dortmund, Germany, under the direction of Hartmut Dieterich who holds the Chair in Vermessungswesen und Bodenordnung.

There is a growing interest in the land and property markets throughout Europe. The Single European Market (SEM) became a reality as early as January 1993, as did the European Economic Area extending SEM benefits to the EFTA countries. Land use and urban development will be influenced by the SEM in many ways, and competition between regions and cities will grow. Furthermore, the Treaty of European Union or Maastricht Treaty, (which explicitly refers to "town and country planning" and "land-use" in Article 130s(2)) is expected to lead to more economic and environmental regulations which will be important factors in the operation of the land market.

The central objective of the EuProMa project was to prepare detailed accounts of the actual operations of the urban land and property markets in five major EC economies. This was necessarily supported by accounts of the planning, taxation and legal framework within which the markets operate.

France, Italy and the United Kingdom were selected because of their importance within the EC, as well as their importance for Germany as trading partners. The Netherlands was of interest because Dutch developers are increasingly active in the German property market, a consideration which is important for Britain also, since many British real estate firms and developers have already set up shop in Germany.

The BMBau commissioned the project because it wanted to know how the land and property market functions elsewhere, why it is as it is, whether the outcome of the market elsewhere is more satisfactory than in Germany, to what extent should changes in the German rules be recommended, and whether new European rules are to be expected or should be proposed. The Ministry was anxious to be able to anticipate any need for new legislation or new policy development, and ensure that in the Single European Market the German development industry was not disadvantaged by any lack of information about how development is undertaken in competitor countries.

It is impossible for one person or one team to gather all the information about land and property in other countries necessary for a well-founded comparison. The co-ordinators in Dortmund were glad and grateful to be able to fall back on the assistance of other members of PRODEST EUROPE, Property Development and Planning Studies in Europe, a consortium of several European universities which is seeking to identify the training requirements of participants in the European property development and planning fields and to promote research and the provision of an educational infrastructure for the development of professional skills in this field. Country reports were completed in the University of Newcastle upon Tyne for the UK, the University of Nijmegen for the Netherlands, the École Polytechnique in Paris for France and the University and Polytechnic of Turin for Italy. Sweden is not yet a member of the EC and therefore not part of the EuProMa project, but a book for the series, adopting the same framework and structure, has been produced by the Royal Institute of Technology, Stockholm. The invitation was extended to Sweden in view of the agreement between the EC and EFTA to form the European Economic Area. There is also clear evidence that the Swedish property and real estate industry and profession is already operating on a European basis.

English was the working language of the different national teams and it was agreed that the national reports were to be written in English or (with the exception of the British report) in "Euro-English", the modern equivalent of the Latin of the Middle Ages. Therefore, English was also agreed to be the language of the publication. The task of series editing was shared between the project director and the UK members of the team, the latter bearing the much larger share of the burden.

To use the books prudently the reader should know a little about the methodological path of the whole project.

In order not to get lost among the many different specialist aspects of the task, each team operated on the basis of the same detailed examination pattern prepared by the co-ordinators in consultation with the other contributors. This pattern forms the basis of the chapter structure of each of the volumes in the series. Thus, comparability between the national studies is ensured.

The separation of land and property (Parts II and III of each book) may require some explanation. Although in most countries the great majority of the material is equally applicable to both, the two markets often do have their own operating characteristics. However, in discussion with the whole team when the framework was being formulated, it became clear that in the case of the Netherlands there were two quite different sets of procedures and actors, and that the best way to expose this on a comparative basis was by this separation in the examination pattern.

The operation of the land and the property market of any country is also determined by general economic, political and cultural conditions in the respective country, the constitutional and legal framework and the rôle and independence of local authorities. Factors such as forms of tenure, macroeconomic variables, performance of the economy, social changes, demographic development, owner-occupation, the requirements for land and the trends in spatial development had all therefore to be taken into account and form the subject matter of Part I of each book. The land market is being influenced, for example, by the conditions for and the process of land-use planning, the process of land assembly (including the development process), the process of construction and the regulations governing the first use of buildings. The level of detail to which the teams were required to work can be seen by looking at the following excerpt of the examination pattern for the framework for the urban land market.

SECTION II: THE URBAN LAND MARKET

1. *The framework within which the market of urban land functions*

1.1 *The legal environment*
(The legal, esp. town-planning system as it affects the conversion of land to the first urban use.)

1.1.1 Law - Acts - Competences - Plans

- Hierarchy of competences
(Hierarchie/Zuständigkeiten)
- Planning acts and their hierarchy
(Gesetzeshierarchie)
- Plans (application, importance)
(Pläne; Anwendung, Bedeutung)
- Presentation in a diagram
(Darstellung in Schemata)
- Obligation/binding character of the plans
(Verbindlichkeit der Pläne)
- Possibilities for higher tiers of administration to intervene
(Eingriffsmöglichkeiten der höheren Ebenen)
(. . .)
- Environmental protection laws
(Umweltrecht)
- Landscape protection (Landschaftsschutz)
- Water protection (Wasserschutz)
- Law of environmental impact assessment (EEC/85/337)
(Umweltverträglichkeitsprüfung)
- Trade inspection/Nuisance
(Gewerbeaufsicht/Immissionsschutz)
- Other relevant acts
(air, noise, public traffic, housing, etc.)
(Sonstige relevante Nebenrechte - Luft, Lärm, ÖPNV, Wohnungsbau, etc.)

1.1.2 Planning process for local plans

- Planning process (formal) for local plans, which confer the right to construct a building
(lokale Ebene für Pläne mit Baurechtscharakter)
- Participation of the citizens
(Bürgerbeteiligung)

- Duration of the planning process (Dauer des Planaufstellung)
- Legal process and administrative court proceedings (Rechtsweg und Verwaltungsgerichtsbarkeit)
- Modification of local plans (Änderung von Plänen)

1.1.3 Private law relating to land, land transactions and first urban use

- Real estate contract (Grundstücksvertrag)
- Servitudes (Dienstbarkeiten)
- Subdivision of plots (Grundstücksteilung)

1.1.4 Instruments for the implementation of plans

- Private law
- Public law (Öffentlich-rechtlicher Art)
- Replotting of land (Bodenordnungsinstrumente)
- Instruments to protect planning, e.g. development freeze (Plansicherungsinstrumente)
- Compulsory purchase/Compensation (Enteignung/Entschädigung)
- Improvements and their charges (Erschließung, -skosten)
- Planning permission (Bauordnungsrecht)
- Betterment levies for planning, building-land regulation, redevelopment (Abschöpfung von Wertsteigerungen bei Planung, Bodenordnung, Sanierung)
- Order to construct (Baugebot)
- Right of first refusal (Vorkaufsrecht)

1.1.5 Information systems

- Differentiation in private and public systems
- Land register (Grundbuch)
- Cadastre (Kataster)
- Valuation committees (Gutachterausschüße)
- Building-land information systems (Baulandinformationssysteme)
- Census (Volkszählungen)

1.2 *The financial environment*
(Who finances and under what conditions)

1.2.1 In general
(. . .)

- Credit practice (credit securities, amortisation/repayment, interest rates (Kreditsystem: Sicherheiten, Zinsniveau, Tilgung)

- Restrictions on capital import (Beschränkungen für ausländisches Kapital)

1.2.2 Per sector
- Possibilities of financing (Finanzierungsmöglichkeiten)
- Investors (Investoren)

1.2.3 Land banking (Bodenvorratspolitik)

1.2.4 Transaction costs

1.3 *Tax and subsidy environment*

1.3.1 Taxes concerning the land market
- Taxation of land (Besteuerung von Grund und Boden)
- Taxation of other kinds of capital/assets (savings deposits, shares, capital income tax, . . .) (Besteuerung anderer Vermögensarten)
- VAT – value added tax (Mehrwertsteuer)

To understand and to evaluate the complicated land and property markets, three components have to be distinguished: (a) the framework within the market is functioning; (b) the process and interaction between different actors who, while operating within the same framework, have different aims and are playing certain specified rôles; (c) the outcome of the market, or the result of its operation. As a fourth and final step, the outcome, as well as the framework and the process leading to the outcome, all have to be evaluated.

Case studies were a central component of the project, six being provided from each country. These are essentially illustrations of the normal operations of the market. They are representative and not special cases, selected on the basis that they should illustrate different aspects of the functioning of the market and the development process in the respective country.

Clearly it was necessary when working on land and property markets to agree on criteria for judging the market, its performance and its outcome. We agreed to use as a basis for this the formula put forward by Hooper for the OECD Urban Affairs Programme:

> The general aim (of the land market) is usually to secure that development land is supplied in needed quantities, appropriate locations, appropriate tenure, at the right time and at appropriate prices, having regard to issues of economic efficiency and social equity. (Hooper, A., "Policy innovation and urban land markets", OECD Urban Affairs Programme, Paris 1989: 5)

It was for the national teams to use this formula to derive the criteria by which to judge the performance and the outcome of their land and property markets, including considerations of social equity and ecological aspects.

Having completed the national research reports, it seemed to all participants and contributors that the material assembled was too valuable to leave unpublished and only referred to in the final comparative study of the German team. The opportunity could not be missed to make these comprehensive studies about the land and property markets of major European countries available to a wider public. The publication of this book is the last in the series. It is a fortunate coincidence that it appears now, after the enormous and radical changes that have taken place in Italy, for the author has updated the entire text to provide a clear analysis of the current (1996) situation.

As series editors we would like to thank all authors of the national reports and books for their support and co-operation in the publishing programme after they had thought they had finished all their hard work by completing the research reports. We would also like to thank the German Ministry for Regional Planning, Building and Town Planning for the generous agreement that the material assembled in reports commissioned by the Ministry and sponsored from its research budget can be disseminated to reach a wider audience in this way.

HARTMUT DIETERICH RICHARD WILLIAMS BARRY WOOD
Dortmund Newcastle upon Tyne

SEPTEMBER 1996

PREFACE

This book is a voyage in the planning system and in the property markets of Italy. I hope it will help the readers to see Italy in a new way, because a real voyage of discovery (as Marcel Proust maintained) is not finding new lands, but seeing with new eyes.

When I started to write this book I thought that it could be a straightforward update of the report on Italy that I edited in at *COREP* (Consortium for research and continuous education at the Polytechnic of Turin) in 1991. The report was produced in the framework of the EuProMa Project, which covered urban planning systems and property markets in five European countries and that had been promoted by the Faculty of Spatial Planning of the University of Dortmund. I soon had to discard that idea because in the course of the 1990s Italy had changed so much as to require an entirely new approach. Thus, I wrote a brand new text between January 1995 and July 1996 on the basis of the most recent facts and data, along the same lines of the original EuProMa Project to ensure comparability with the other books in the series. Most of the tables and graphs reflect the situation at the end of 1995 or in the first quarter of 1996.

Writing this book would have not been possible without the help of many people to whom I am most grateful. Some have been so important to me that I must mention them, without giving to them any responsibility for any inaccuracies in the text. First of all, I thank Paolo Ceccarelli, Dean of the School of Architecture (*Facoltà di Architettura*) of the University of Ferrara. He read the final version of the text and provided useful advice which I hope to have utilized well. If the book contains any innovative ideas on city management and on the relationship between city master plans and local policies for the retail sector, this is because since 1993 I have been involved with several research projects directed by Paolo Ceccarelli at the *URB* (Center for Urban Research) of the University of Ferrara.

I must thank Franco Corsico and Luigi Mazza (respectively at the Polytechnics of Turin and Milan). I have been working for years with Franco Corsico on the issue of urban marketing and industrial wasteland, and most of my best ideas on these subjects, as shown in this book, derive from this association. Likewise, what I have written on the issue of urban planning and sustainable growth comes from my participation in an international research programme directed by Luigi Mazza and financed by DG XII of the European Commission within the framework of the Environment Research Programme between 1992 and 1994. I thank also Mario Rey (University of Turin), Giuseppe Dematteis and Alberigo Zeppetella (Polytechnic of Turin) whose contributions to the report of 1991 have been a valuable reference point for my ideas on the issues of urban trends, real estate financing and environment policies.

The case studies are based on first-hand accounts and documents provided to me by most of the actors involved. To this end, I wish to thank all those who spent time and effort in the following institutions and firms: Commune of Beinasco, Commune of Turin, *Coop Piemonte*, *Gabetti Holding* (Milan) and *Gabetti* (Genoa), *MEDEDIL-Società Edilizia Mediterranea*, *Milano Centrale Immobiliare* (Gruppo Pirelli), *Sistemi Urbani* (Milan).

I am grateful to Elena Molignoni and Stefano Stanzani at Nomisma research institute in Bologna, for their invaluable assistance in finding the most accurate data on the property markets. Cesare Paonessa, a graduate of *COREP* and now a freelance consultant, provided a significant contribution in updating several tables. He also helped me with document gathering and analysis for one of the case studies. Likewise, the personnel of the *URB* (Centre for Urban Research) in the *Facoltà di Architettura* at the University of Ferrara provided overall assistance. Daniela Mascellani and Francesca Lorenzetto helped in drafting the maps of the case studies.

The text, which I wrote directly in English, has been checked for style consistency with the other titles in the series by Barry Wood and Richard Williams. To Barry Wood I send special thanks for all his assistance and encouragement. At his suggestion, to facilitate understanding of the intricacies of urban planning and property markets in Italy, I have included a summary of the main legislative acts. For the typing there is no one to thank but WordPerfect and my portable computer which accompanies me in my weekly commuting between Turin and Ferrara.

Finally, I am sincerely grateful to Rosy, my companion, for her support and understanding, in spite of all the time that this book demanded.

GASTONE AVE
FERRARA, SEPTEMBER 1996

Abbreviations and acronyms

ABI	*Associazione Bancaria Italiana* (Association of Italian banks)
AC	amount of compensation
AEM	*Azienda Elettrica Municipale* (Municipal Electricity Company)
ALFA	*Anonima Lombarda Fabbrica Automobili*
ANAS	*Azienda Nazionale Autonoma delle Strade* (independent national road company)
ANCE	*Associazione Nazionale Costruttori Edili* (National Association of Construction Companies)
APT	*Azienda di Promozione Turistica* (Agency for Tourism Promotion)
ASCOM	*Associazone del Commercio del Turismo e del Servizi* (Association for Commerce, Tourism and Services)
ATC	*Agenzia Territoriale per la Casa* (Local Agency for Housing)
AUDIS	*Associazone delle Aree Urbane Industriali Dismesse* (Association of Vacant Industrial Areas)
AV	actual market value of the property
BNL	*Banca nazionale del lavoro* (National Bank of Labour)
BOC	*Buoni ordinari comunali* (municipal bond)
CAAT	*Centro Agro Alimentare Torino* (freight platform for perishable goods, Turin)
CCDC	Centre City Development Corporation
CELP	Centre for Environmental Law and Policy
CER	*Comitato per l'Edilizia Residenziale* (Committee for Housing Programmes)
CENSIS	*Centro studi investimenti sociali* (Centre for Social Studies)
CGIL	*Confederazione Generale Italiana Lavoratori* (Italian General Confederation of Workers)
CIPE	*Comitato interministeriale prezzi* (Inter-ministerial Committee for Prices)
CISL	*Confederazione Italiana Sindacati Lavoratori* (Italian Confederation of Worker's Unions)
CIT	*Compagnia Italiana Turismo* (Italian Tourism Company)
CNR	*Centro Nazionale delle Ricerche* (National Research Centre)
COMIT	*Banca Commerciale Italiana* (Italian Commercial Bank)
CORECO	*Comitato Regionale di Controllo* (Regional supervising committee)
COREP	*Consorzio per la Ricerca e l'Educazione Permanente presso il Politecnico di Torino* (Consortium for Research and Continous Education at the Polytechnic of Turin)

CPI consumer price index
CR rent value as given by the cadastre books
CREDIT *Credito Italiano* (Italian Loan Bank)
CRESME *Centro Ricerche Economiche, Sociologiche e di Mercato nell'Edilizia* (Centre for Economics, Sociology and Marketing Research in Real Estate)
EC European Community
ECU European Currency Unit
EMS European Monetary System
ENEA *Ente per le Nuove Tecnologie, l'Energia e l'Ambiente* (Agency for New Technologies, Energy and Environment)
ENEL *Ente nazionale per l'Energia Elettrica* (National agency for electric energy)
ENI *Ente Nazionale Idrocarburi* (National Oil Agency)
EU European Union
FAL *Foglio Annunci Legali* (Register of Legal Acts)
FIAP *Federazione Italiana Agenti Immobilari Professionisti* (Federation of Italian Professional Property Agents)
FIAT *Fabbrica Italiana Automobili Torino*
FIMAI *Federazione Italiana Mediatori e Agenti Immobiliari* (Italian Federation of Real Estate Agents)
FIO *Fondo per gli Investimenti e l'Occupazione* (Fund for Investments and Employment)
IMF International Monetary Fund
FS *Ferrovie dello Stato* (Italian state railway company)
GDP Gross domestic product
GESCAL *Gestione Case Lavoratori* (Fund for Housing of Employees)
GMA Greatest metropolitan area
GNP Gross national product
IACP *Istituto Autonomo Case Popolari* (self-governing housing association)
ICF *Istituto di Credito Fondiario* (real estate long term financing institutions)
ICI *Imposta comunale sugli immobili* (local tax on real estate properties)
ICIAP *Imposta comunale industria arti e professioni* (local tax on business activities)
INA *Istituto Nazionale Assicurazioni* (National Insurance Company)
INA-CASA *Piano per le abitazioni dell'Istituto nazionale assicurazioni* (Housing Programme of the National Insurance Company)
INAIL *Istituto Nazionale per le Assicurazioni contro gli Infortuni sul Lavoro* (National Insurance Company for Labour Accidents)
INPDAP *Istituto Nazionale di Previdenza per i Dipendenti dell'Amministrazione Pubblica* (National Insurance Company for Public Employees)
INVIM *Imposta sull'incremento di valore degli immobili* (tax on accrued value of real estate properties)
INU *Istituto Nazionale di Urbanistica* (National Institute of City Planning)

IPI — *Istituto Piemontese Immobiliare* (Real Estate Company of Piedmont)
IRI — *Istituto per la Ricostruzione Industriale* (national holding for industrial reconstruction)
IRPEF — *Imposta sul reddito delle persone fisiche* (individual income tax
IRPEG — *Imposta sul reddito delle persone giuridiche* (income tax of companies)
IRS — *Istituto Ricerche Sociali* (Institute for Social Research)
ISCO — *Istituto Nazionale per lo Studio della Congiuntura* (National institute for the Study of the Economy)
ISTAT — *Istituto Centrale di Statistica* (National Bureau for Statistics)
ITALGAS — *Società Italiana Agenti Immobiliari Professionisti* (Federation of Italian Professional Property Agents)
IVA — *Imposta sul valore aggiunto* (value added tax)
MEDEDIL — *Società Edilizia Mediterranea* (real estate company)
NCEU — *Nuovo Catasto Edilizio Urbano* (New Urban Cadastral Bureau)
PCM — *Piano di comunità montana* (Mountain Community Plan)
PEEP — *Piano di edilizia economica e popolare* (Low-cost Housing Plan)
PF — *Piano di fabbricazione* (development plan)
PIM — *Piano Intercomunale Milanese* (Milan inter-communes Master Plan)
PIP — *Piano insediamenti produttivi* (plan for light industrial development)
PL — *Piano di lottizzazione* (allotment plan)
PP — *Piano particolareggiato* (detailed plan)
PPA — *Programma pluriennale di attuazione* (multi-year implementation programme)
PR — *Piano di recupero* (urban renewal plan)
PRG — *Piano regolatore generale* (city master plan)
PRGC — *Piano regolatore generale comunale* (see PRG)
PRGI — *Piano regolatore generale intercomunale* (metropolitan area master plan)
PTC — *Piano territoriale di coordinamento* (plan for territorial coordination)
PTI — *Piano territoriale infraregionale* (territorial plan over two or more regions)
PTP — *Piano territoriale paesaggistico* (landscape plan)
PTR — *Piano territoriale regionale* (regional territorial plan)
PUP — *Piano urbano dei parcheggi* (urban plan for car parking)
PUT — *Piano urbano del traffico* (urban plan for traffic)
PZ — *Piano di zona* (local plan)
RAI — *Radio televisione italiana* (Italian state radio and television company)
SAS — *Società in accomandita semplice* (personal responsibility company)
SDO — *Sistema Direzionale Orientale* (eastern office areas)
SPA — *Società per azioni* (stock company)
SRL — *Società a responsabilità limitata* (limited company)
STET — *Società per la Telefonia e le Telecomunicazioni* (Telephone and Telecommunication Company)
TAR — *Tribunale amministrativo regionale* (regional administrative court)

TAV	*Treno Alta Velocità* (high-speed train)
UIL	*Unione Italiana del Lavoro* (Italian Union of Labour)
URB	*Centro Ricerche Urbane della Facoltà di Architettura di Ferrara* (Centre for Urban Research, School of Architecture, University of Ferrara)
USSL	*Unità socio sanitaria locale* (local health service)
UTE	*Ufficio tecnico erariale* (local bureau for technical appraisal)
VAT	value added tax (see IVA)
WHO	World Health Organization
WPS	wider public sector
ZTL	*Zona a traffico limitato* (restricted traffic area)

Preface to the 2018 edition

Italy has gone through many social and economic changes since the first edition of this book in September 1996. However, the main social and economic structures are still basically the same. Thus, this book remains a good starting point to explore the intricacies of Italian urban planning and real estate development practices. Yet, the reader should be aware about the main changes which have affected the scenario of land and property markets in contemporary Italy, as briefly outlined below.

Shift from Italian lira to euro, the single European currency unit.

The Maastricht Treaty, signed in 1992, set the agenda to create the economic and monetary Union of European states. On the 4th January 1999, prime minister Romano Prodi announced that Italy was to adopt the single European currency, the euro, starting on 1st January 2002, together with a first group of 10 other European nations (Austria, Belgium, Finland, France, Germany, Ireland, Luxemburg, Netherlands, Portugal, and Spain). The exchange rate was fixed at EUR 1.0 = ITL 1,936.27. On 1st January 2002, the first 11 countries were joined by Greece, which had been admitted in 2001.

The euro has become, since its introduction, the most used currency for foreign trade transactions in the world after the US dollar, as shown by the *Triennial Central bank Survey* of the BIS-Bank for International Settlements (see: https://www.bis.org/publ/rpfx16.htm , last visited 10 January 2018). The Japanese yen ranks in a distant third place on one side of world trades.

As of 2018, the euro area comprises 19 nations, the initial 12 counties mentioned above plus Cyprus, Estonia, Latvia, Lithuania, Malta, Slovakia, and Slovenia. In addition, the euro is used as national currency by unilateral choice in some small European states which are not part of the European Union (e.g. Montenegro). Today, more than 340 million people live in the euro area. Within a decade the number will exceed 500 million people, as other nations are in line to adopt the new currency (e.g. Bulgaria, Croatia, Czech Republic, Hungary, Poland, and Romania).

Finance and the real estate industry

The shift from the Italian lira to the euro had two major effects on the property markets in Italy. Firstly, the cost of real estate loans dropped from the high rates of the nineties (in the range of 10-15 percent for fixed rate loans) to a rate of 4-5 percent at the entrance of the euro, and to a rate of just 1 percent in the current decade. This change pushed up the demand for new loans starting in 1999 as they were denominated in euro, even before its material circulation in 2002. The bullish market led to the record high number of 869,000 units sold in 2006, compared to an average of about 500,000 home units sold in previous years. Secondly, the Italian real estate industry opened up to foreign investors. All risks of devaluation of the Italian lira against other major currencies vanished with the euro to replace the national currency.

Birth and development of the Italian real estate closed-end funds.

It is by no chance that the first Italian real estate fund (*Valore Immobiliare Globale*) was launched on 15th February 1999, followed in the same year by two other funds (*Securfondo, Unicredito Immobiliare Uno*). These were retail funds approved by the Bank of Italy as investment vehicles for the general public, and traded on the stock exchange. In a decade about 30 retail funds hit the market, while private real estate funds had an even greater growth.

By the end of 2017, there were in Italy 407 real estate funds, of which 18 were retail funds (Scenari Immobiliari, *I fondi immobiliari in Italia e all'estero. Rapporto 2017*, Roma, 2017, p. 265). Currently, the net asset value (NAV) of properties managed by Italian real estate funds, is estimated at 50 billion euros (Scenari Immobiliari, *I fondi immobiliari in Italia e all'estero. Aggiornamento Rapporto 2017*, Roma, 2017, p. 8). Italian funds rank in fourth place in Europe, together with Luxemburg funds, after Germany (164 billion euros), France (127 billion euros), and Netherlands (70 billion euros).

The success of real estate funds renovated deeply the real estate industry in Italy. As movers and shakers of the property market, they put on top of their search lists buildings with high architectural quality and energy efficiency standards, not just with prime locations. Developers had to follow suit, so the quality of new buildings greatly improved along the past 20 years. The new funds contributed to a spread in Italy of a new professional culture, including the technique of property appraisal based on the discounted cash flow (DCF) analysis. The method is not flawless if it is applied mechanically, without due knowledge to real estate and planning issues. The DCF analysis should be just a component of a complete evaluation, including legal and urban planning reviews with thorough site visits.

The Italian real estate funds centered their activities in office and retail properties located in high street locations in Rome and Milan. Overtime the search focus was enlarged to other property sectors (e.g. student housing buildings) and to other Italian cities, but only rarely the funds invested in cities in the South and in the islands of Sicily and Sardinia. These areas may become the next new frontier of real estate investments in Italy.

New tax deductions for housing renovations, energy savings and seismic safety upgrades.

In 1998, during the first government led by Prodi, a tax deduction of 41 percent for housing renovation works according to a detailed set of rules, was introduced for the first time. The rationale was to fight tax evasion, as the deduction could be obtained only for invoices paid through bank draft. It was a clear success and it was renovated year by year until it was made permanent in 2012. The percentage allowed for tax deduction has changed several times in the period 1998-2017. In 2018 it is 50 percent, and it should remain constant for the coming years. This measure was a policy choice in favor of the reuse and maintenance of the existing housing stock, as opposed to new buildings. It helped greatly to curb new greenfield projects and urban sprawl.

In 2007, during the second government led by Prodi, a tax deduction of 55 percent for renovation works and equipment (heating units, air conditioners etc.), apt to increase energy savings of real estate properties, was introduced. The percentage was soon raised to the current level of 65 percent. This measure has been a success even greater than the incentive for renovation works. In 2013, a tax deduction of 65 percent on works for seismic safety of existent homes and other buildings was introduced, with maximum expenditure of 96,000 euros per unit, spread over 10 years fiscal instalments. The state budget law for 2018 raised tax deductions to 85 percent for works which make a building 2 steps safer in a national seismic scale. This measure will remain in effect until the end of 2021.

These new tax deductions are part of an array of other fiscal measures (e.g. tax incentives for new appliances and furniture in renovated homes; incentives for renovated multifamily buildings) which cover a time lapse longer than 20 years, and make Italy probably the best example in Europe for long-term policies in favor of maintenance of the existing housing stock. This provides a positive effect on restraining urban expansion. This achievement was obtained with the support of all economic players involved, including the main association of building companies ANCE (Associazione Nazionale Costruttori Edili) whose past president, the late Claudio De Albertis, contributed to the renovation of the real estate industry in Italy.

Urban strategic planning makes its way into the ordinary planning system.

Turin, the third largest city in Italy and the first national capital in 1861, was the first city in Italy to approve an urban strategic plan, on the 29th February 2000. In May 1998 Valentino Castellani, mayor of Turin and head of the Development Forum, a public-private partnership including surrounding cities, announced the decision to prepare the plan. It was completed in June 1999 when the International Olympic Committee, gathered in Seoul (South Korea), selected Turin as the venue of the XX Olympic Winter Games scheduled in 2006. This led to a full rewriting of the plan in a record time. The plan, and the event *Torino 2006,* with its urban infrastructures (e.g. the underground), were a tool to transform Turin from an industrial city, to a post-industrial metropolis.

The planning team was led by urban economist Roberto Camagni, coordinator of the scientific committee, and by sociologist Arnarldo Bagnasco. The team benefitted of the knowledge and passion of the late Franco Corsico, city commissioner for urban planning. I was part of the team as urban planner from the start, and was in charge of urban planning issues including the coordination of the new strategic approach with ordinary planning procedures.

Thus, the plan was the product of a truly interdisciplinary team, and was made possible by the support of a public-private partnership. The goal was to identify a new long-term vision for the metropolitan area, to transform the city from a one company town to a balanced international city - as the title "*Torino Internazionale*" made clear. The plan did not concern land use regulations, as it was not intended to replace the city master plan, which had been approved in 1995. The new plan consisted of 84 tangible and intangible projects in different sectors (e.g. local government, culture, tourism, sports, urban infrastructures, etc.). The plan was successfully implemented through a governance approach to planning and a joint responsibility between public and private players.

The strategic plan of Turin is recognized as a reference planning tool, and it sets a divide line in recent urban planning in Italy. This is shown by the most comprehensive Italian text on urban strategic planning - A. Tanese, E. Di Filippo, R. Rennie, editors, La pianificazione strategica per lo sviluppo dei territori, Rubbettino Editore, 2006. Soon after its approval, many other Italian cities tried to approve and implement their own strategic plan. Only some cities were successful (e.g. Trento, Genoa, Perugia). Nevertheless, the method of urban strategic planning is here to stay. Unsurprisingly, the new planning method has inspired more urban policy makers and local government officials than the academia. There are odd confusions between the two kinds of plans, urban strategic plans on one side, city master plans on the other side. Yet, many Italian cities look forward to developing a full merge between land use plans and strategic plans.

An issue which awaits action: urban planning as a public service.

This book lists at the end five issues awaiting action. The first topic is the fiscal system, whose main limits have been well addressed since 1996, as outlined above. The second is how Italian cities are now performing better on linking land use and transport planning, as well as on the third issue, reducing red tape. The last two issues deal with timing of planning procedures, and implementation capability. Here there is still much to do. The task relies upon both local governments and private companies, with the increased need for public-private partnerships.

In fact, the history of land and property markets in Italy can be read as a continuous fight between those who want to serve the public interest, and those who care for their wealth only. I include in the first group the most advanced players of the real estate industry, while in the second group, there are those who still rely for their fortunes on land use changes and windfall gains generated by taxpayers money that is then spent on mass transport lines and other public infrastructures.

Urban planning remains in Italy, an essential public service to restrain the misuse of land, prevent urban sprawl, and do something positive for our little planet.

Gastone Ave

Ferrara, January 2018.

PART I
Overview of the main conditions in Italy

CHAPTER 1
Basic information and connections

1.1 The constitutional and legal framework

Constitution and organization of the state

Italy is a republic whose constitution came into force on 1 January 1948. The constitution defines value systems, the main principles of power-sharing, the rules for drawing up regulations, and the distribution of responsibilities between the different levels of government. Laws revising the constitution have to be approved by both houses of parliament, with an absolute majority in each, following two successive debates within a period of not less than three months.

Legislative power is the prerogative of the two branches of Parliament, House and Senate. There is no particular function that distinguishes them, as the text of a proposed law must be approved by both in identical form. The power to enact subsidiary legislation is delegated to the government in respect of "delegation laws", which must be confined to limited purposes. Laws regarding constitutional matters, budgets, statements of account and authorization of provisional procedures cannot be delegated.

The constitution allows for the possibility of issuing decrees directly by the government only in cases of extreme necessity and urgency. These have to be submitted to the House on the day of issue in order to be made law. The time limit for ratification is 60 days, otherwise a decree has to be withdrawn. Issuing amendments identical to those not ratified is prohibited. Nevertheless, both this regulation and the requirement for "necessity and urgency" now tend to be interpreted in a rather liberal fashion. This tends to strengthen executive power.

The President of the Republic is elected by both branches of Parliament in the same sitting. His powers are very limited. Essentially, they consist of the naming of the President of the Cabinet, the proclamation of laws (an obligatory act, except for the power to call for their review), the exercise of the functions of President of the Senior Council of Magistrates, and the role of head of the armed forces.

It should be noted, however, that the President of the Republic has assumed a more substantial role in the 1990s, in the power vacuum created by the decline and dismantling of the main political parties (Christian Democrats and Socialists). For example, President Oscar Luigi Scalfaro played a clear political role in steering the country towards the general elections of 1996 after the crisis caused by the fall of the government led by Lamberto Dini and the ensuing stalemate between the centre-right and centre-left coalitions.

The constitution guarantees the autonomy of the judiciary (penal, civil and administrative), laying down a series of general principles of independence permanence, absence of hierarchy and instituting the Senior Council of Magistrates as a self-governing body. It is up to the Constitutional Court to guarantee both the constitution and balance of power. Within its control fall disputes between regional and national laws, conflicts of responsibility between the powers of the state and the regions, and between the different regions, monitoring the activities of the President and ministers of the Republic. One third of the members of the Court are named by the President of the Republic, another third by the Houses and the final third by the supreme members of the judiciary.

The electoral system in force is based, with some exceptions (e.g. the Senate, provincial courts and communal councils with jurisdiction over 10000 inhabitants or fewer) on the principle of proportional representation. This leads to a greater number of parties than is usually found in other European countries, some of which only have 1% of the votes. Many people are of the opinion that the Italian system of proportional representation has some negative effects: fragmentation of the representation of interests, difficulty in the approval of executive laws, disparity between electoral weight and institutional weight, and situations frequently occurring of conflicting or "crossed-vetoes" in very important decisions. Accordingly, different proposals to modify the electoral system have recently been proposed. A new electoral law was passed in 1993.

This reform reduced the total number of members of Parliament elected via the proportional system to 25% and introduced the majority system, according to which the winning coalition would receive a premium in terms of Parliamentary seats, in order to ensure a higher degree of stability of government. However, the new law is not enough by itself to ensure a stable government for the country. Italian society is searching for a new equilibrium after the breakdown of the traditional political parties, particularly the Christian Democrats, which had been in power with no interludes for over four decades. The Christian Democrats and the Socialists, together with minor parties (including Social Democrats, Republicans and Liberals), which were part of the ruling coalitions until the early 1990s, were literally eliminated from the Italian political scene by the discovery of the system of corruption that had been utilized for decades in order to finance the political parties and their key leaders.

Since the electoral law of 1993, one quarter of the members of Parliament are elected according to the previous proportional system, and the rest according to a non-proportional system, which increases the likelihood of a party having an overall majority. At the local level, mayors are elected directly. Whenever one candidate has over 50% of the vote in the first round, that person is elected. Otherwise, the top two candidates from the first round contest a second round. Previously, mayors were elected by the parties who formed the winning coalition, and all seats assigned on a strictly proportional basis. The fragmented representation resulting from this was cited as the cause of unstable local government, in which the mayors tended to be more responsive to their parties than to the citizens at large. At present, the coalition that supports the winning mayor candidate is given at least 60% of the seats in the commune council in order to guarantee stable local government for the full length of its four-year term of office.

Decentralization of powers

Regions Regional constitutions represent a clean break with the centralization of fascist Italy. In the years immediately following the end of the Second World War, five regions were instituted by special statute and were accorded a large degree of autonomy. These five are the Aosta Valley, Friuli-Venezia Giulia, Trentino-Alto Adige, Sicily and Sardinia. However, regions with ordinary statutes were only instituted in 1970. In total, there are twenty regions, eight in the North of the country, four in the Centre, six in the South and the two islands of Sicily and Sardinia. Italy has been unified since 1861, and the terms north, centre and south are officially only of geographical significance. The regions correspond, more or less, to the various autonomous states of the Italian peninsula before the unification of the country, and even maintain a link with ancient Rome, when, for example, under the Emperor Augustus there were eleven regions, some with exactly the same names and boundaries as today's regions. The Italian regions (*regioni*) and their capital cities (*capoluoghi di regione*) are listed in Table 1.1.

Table 1.1 The regions of Italy and their capital cities.

Region	Main city	Geographical location
Aosta Valley	Aosta	North
Piedmont	Turin	North
Lombardy	Milan	North
Veneto	Venice	North
Trentino–Alto Adige	Trento	North
Friuli–Venezia Giulia	Trieste	North
Emilia–Romagna	Bologna	North
Liguria	Genoa	North
Tuscany	Florence	Centre
Umbria	Perugia	Centre
Marche	Ancona	Centre
Lazio	Rome	Centre
Abruzzo	L'Aquila	South
Molise	Campobasso	South
Campania	Naples	South
Puglia	Bari	South
Basilicata	Potenza	South
Calabria	Catanzaro	South
Sicily	Palermo	Islands
Sardinia	Cagliari	Islands

Regions have the power to issue laws within their own boundaries, and these are respected by the Constitution and constitutional law. They have to be flexible in order to operate within the limits of the fundamental principles of any national law that applies to the same areas, and cannot be in conflict with the interests of the nation or other regions. Regional authorities have legislative powers over the communes, local urban and rural police, fairs and markets, public charitable institutions (including sanitary and hospital care), museums and local libraries, town planning, tourism and the hotel industry, public transport within the region, navigation and seaports, mineral and natural water, quarries and mining, hunting, fishing in inland waters, forestry and agriculture, and crafts. It is possible to delegate other areas of responsibility to the regional legislatures.

Regions have administrative powers in all the areas for which they have legislative responsibilities, except for those delegated by law to the provinces and communes. Other administrative functions can be delegated by the state to the regions. These in turn can be subdelegated (as in fact often happens) for administrative reasons to the provinces and communes. Regional statutes have a fundamental role in financial and land-use planning. Linking regional development plans and implementation of urban and regional laws have been central to the first two legislatures in most regions. The changed political and cultural climate in the 1980s has tended to relegate this plan to second place, in favour of an increased tendency towards mundane management, ineffectual administration, and the slow distribution of resources. This is also the cause of the fiscal crisis, which is worsening all the time. The Council is made up of regional members elected every five years who exert legislative power, together with the Board (the executive body), and the President (who represents the region). Between them, these three bodies administer laws and other business delegated by the state.

Provinces The principal functions of the provinces include construction and maintenance of provincial roads, contributions to planning and execution of various types of public works (waterworks, ports, etc.), psychiatric help, participation through their institutions, financial contributions and/or consortia towards various services such as preventive hygienic–sanitary treatment, charity, and technical and professional instruction. The provinces had only a limited range of functions until their role was greatly enhanced by the Law of Reform of Local Government Functions (1990). Table 1.2 indicates the total number of provinces (103) and communes (8103) in Italy at the end of 1993. As of 1995 the situation was unchanged, except the number of provinces, which had increased by one to 104. It must be noted that only 10 of the 20 regions have a population of over 2 million people. Similarly, 4684 communes out of a total of 8103 have fewer than 3000 people, and only four communes (Turin, Milan, Rome, Naples) have more than 1 million inhabitants, if we consider just the population inside the administrative boundaries of the communes.

Communes The commune is the basic administrative unit. Its responsibilities can be separated into its own immediate concerns, and those delegated by the state or the regions, both obligatory and optional. Among its responsibilities are those of local policing, health and hygiene, urban cleanliness, charities and benefits, creation of general facilities within the town, granting building licences and setting up public services.

The reform of local finance introduced in 1971 broadened the scope of taxation and financial autonomy for the communes. A series of amendments issued from 1977 onwards put limits on the levels of appropriation and growth of communal

Table 1.2 Number of provinces and communes, by region, as of the end of 1993.

Region	Number of provinces	Number of communes
Aosta Valley	1	74
Piedmont	8	1209
Lombardy	11	1546
Veneto	7	582
Trentino-Alto Adige	2	339
Friuli–Venezia Giulia	4	219
Emilia–Romagna	9	341
Liguria	4	235
Tuscany	10	287
Umbria	2	92
Marche	4	246
Lazio	5	377
Abruzzo	4	305
Molise	2	136
Campania	5	551
Puglia	5	258
Basilicata	2	131
Calabria	5	409
Sicily	9	390
Sardinia	4	376
TOTAL	103	8103

spending. The members of the commune are the Council, the Board and the Mayor whose role has been enhanced by the direct election principles introduced with the electoral law of 1993.

Consortia, mountain communities, and district councils Communes and provinces can organize private contractors to carry out work or manage services for the commune. These contractors, the *Consortia*, are granted a single statute that sets out the aims, administrative and financial quotas. For example, there are consortia for collection and disposal of refuse, and for servicing of refuse vehicles, for land reclamation and for water supplies.

Mountain communities, a designation instituted in 1976, have the main aim of promoting economic development in mountainous areas. They prepare plans for urban communal facilities, indicating the type, locality and foreseeable cost of necessary investment, and creating the necessary incentives. Mountain communities can be subject to delegation or subdelegation on the part of the appropriate region.

The subcommunal decentralization of large communes, which started in 1976, is backed up by the district council. These essentially have to do with convening meetings, formulating non-binding opinions (e.g. on planning arrangements), and presenting proposals related to forecasting balances, services, planning acts and communal laws.

The finances of regions, provinces and communes

Communes Before the reform of local finance came into force in 1973, the communes had considerable powers of taxation: a quota of between 55% and 60% of the current returns. This percentage has now been reduced to 10–15%. Also, the financial community is now based essentially on allocations. In the first period of the reform's application, the lack of allocation of hidden revenues and the length of time it took for this money to find its way into the commune's account gave rise to a rapid increase in borrowing. The situation regarding appropriation tended to become intolerable. In 1977, a series of moderate government laws were introduced as a drastic measure. The calling-in of communal loans was taken on by the Exchequer, but in the meantime further limitations were imposed against the possibility of further borrowing, together with strict limits on the growth of current spending. The clear result is that the autonomy of the communes has been reduced.

Provinces The mechanisms for financing provinces are similar to those of the communes. After the reform of local taxation in 1971, the returns from debts shrank even further than those of the communes. Nowadays, provincial finances are based essentially upon monetary movements. The allocations themselves represent 7–8% of current returns in recent years.

Regions Returns to the regions have been drawn almost entirely from allocations, which have in turn depended on the central budgetary balance of the state (see Table 1.3). Funding is transferred to the regions in the form of the communal fund, the development fund and funding set aside by special decree.

The communal fund comes from fiscal transfers based on fixed quotas. Sixty per cent is divided between the regions proportionally, according to the size of population. Ten per cent is allocated in proportion to the area and 30% is allocated according to an indicator which considers average income, occupation and numbers of migrants. The communal fund is not subject to constraints in expenditure and is the basis on which the overall autonomy of the regions is founded.

The development fund was intended to form the relationship between national and regional planning. The amount of this fund and its distribution ought really to have been determined for each five-year period of the national economic plan, but the mechanism never really worked, because of the progressive abandonment of the ideal of central government planning of the economy. The communal fund is distributed according to similar criteria by those who control the communal funds. A partial renewal of the idea of transfers being related to detailed programmes was set up in 1982 with the creation of the Fund for Investments and Employment (*FIO*).

The *FIO* organization is based at the Ministry of Finance, where a specific office (*Nucleo di valutazione degli investimenti pubblici*) appraises the proposed projects and places them in priority order according to their expected contributions to achieving objectives of national importance. In the first phase of application of *FIO*, most of the funded projects concerned agriculture. Recently, *FIO* has financed

Table 1.3 Allocation from central state to regions, provinces and communes (billions of lire at current prices).

	1980	1981	1982	1983	1984	1985	1986*	1987	1988	1989	1990	1991	1992	1993
Current transfers														
Regions (total)	22622	26640	31101	41769	49166	53601	69957	67401	76397	83820	86452	108447	114032	63796
NHS	18018	19738	23181	31000	35955	38353	40834	46029	53183	58511	64302	79815	82997	36394
All other purposes	4604	6902	7920	10769	13211	15248	29123	21372	23214	25622	22150	28632	31035	27402
Communes & provinces	14125	14501	15413	18726	23099	26170	26773	24223	26010	25007	28349	29886	29384	30271
Capital transfers														
Regions (total)	2926	3800	5839	7881	9455	7620	6377	9524	9010	11815	8357	8870	4769	6902
NHS	164	310	510	701	1094	595	315	979	2591	2688	1101	2654	419	–
All other purposes	2762	3490	5329	7180	8361	7025	6062	8545	6419	9342	7256	6216	4350	6902
Communes & provinces	176	162	234	215	184	1776	9238	10297	13250	12199	10710	14268	10743	10878
Total														
Regions (total)	25548	30440	36940	49650	58621	61221	76334	76925	85407	95635	94809	117317	118801	70698
NHS	18182	20048	23691	31701	37049	38948	41149	47008	55774	61199	65403	82469	83416	36394
All other purposes	7366	10392	13249	17949	21572	22273	35185	29917	29633	34964	29406	34848	35385	34304
Communes & provinces	14301	14663	15647	18941	23283	27946	36011	34520	39260	37206	39059	44154	40127	41149

* Including 14674 billion lire in transfers to Sicily and Sardinia for accounting adjustments, 4479 billion lire concerning previous years.
Source: Ministero del Bilancio.

projects for the restoration of historic buildings and cultural activities, including the computerized census of the artistic properties of all the state museums. The projects can be proposed for funding to the Ministry of Finance by different public bodies, among which are the regions themselves. Funds set aside by special decree are allocated on an ad hoc basis, for example in the case of natural disasters.

Quantitative aspects of regional and local finance

The sizes of the budgets for the regions, provinces and communes are clearly different.The current returns to the regions have fluctuated in the recent years between 7% and 7.5% of the gross national product (GNP), those to the communes between 4% and 4.4% of GNP, and those to the provinces represent only 0.5–0.6% of GNP. Nevertheless, the roles of regional budgets appear to take on new dimensions if the transfers to the health service are taken into account. The latter cannot be amended. Considering the allocations for which they alone are solely responsible, the budget handled by the regions, on the one hand, and the communes and provinces, on the other, are very similar. In recent years, allocations to the capital account are in fact greater for the communes than the provinces. It should be noted that between the end of the 1980s and the first part of the 1990s, both current transfers and capital transfers to the regions stopped growing at the rapid rate of the previous decade. Recently, the main cut was achieved in the National Health Service.

The reform of local government functions

Reform of existing practices governing the activities of the communes and provinces (going back to the reform of 1934) was delayed by political forces and administrative sensibilities. After a long and drawn out debate, this reform was finally made law in June 1990, and it contains important changes. The most important points are reviewed here.

Statutory autonomy Communes and provinces are automatically given the power to modify their own laws, organize themselves and their areas of responsibility, set up their own administration and services, and take measures to ensure public participation and guarantee the right to information. Written statutes dictate the kind of collaboration between communes and provinces, and the regions are expected to establish, with their own laws, general cooperation between representatives of local bodies and the region itself.

Tasks of the commune Communes assume responsibility for social services, land use and local economic development. The commune previously had the autonomous power to determine land use within its territory. The innovation is in the linkage between this role and the brand new task of promoting economic development at the local level. As will be highlighted throughout the text, this is probably the single most important innovation of the Italian planning system, because it will force both local administrators and city planners to establish a sound link between the city master plan and the financial budget plan of each commune.

Tasks of the province The innovations here are radical. Provinces are required to take on the role of intermediaries between communes and regions. They take full responsibility for certain areas. These changes bring together two needs that, in the late 1970s and in the early 1980s, became part of the administrative remit of many Italian regions as part of the process of regional decentralization, namely the tasks of defining politically sensitive land in vast urban areas, and of coordinating initiatives and communal urban facilities. Since provinces had never enjoyed clearly defined administrative responsibilities and therefore possessed neither the usual facilities nor financial autonomy, the experience has so far revealed extensive limitations. The redefinition of the province's functions within a framework of statutory autonomy represents a new opportunity for carrying out policies and plans at the subregional level.

The responsibilities assigned to the province by the new law are to be carried out either in large intercommunal areas or within the province as a whole. They relate to soil conservation, protection and evaluation of the environment and prevention of disasters, protection and evaluation of water and energy resources, evaluation of cultural artefacts, roads and transport, protection of flora and fauna, parks and nature reserves, hunting and fishing, the organization of waste disposal at provincial level, surveys, control of water outlets and of atmospheric and noise pollution, health services, hygiene and preventive health care, linked to secondary education and professional training (which is delegated by the state and regional legislation), collection and processing of data and technical-administrative assistance for local bodies.

Designation of metropolitan cities Another substantial innovation in the law involves setting up a form of metropolitan government for areas based on the nine major cities of Turin, Milan, Genoa, Venice, Bologna, Florence, Rome, Naples and Bari, and their surroundings. The functions allocated to the provinces nationally are delegated to them. It falls to the regions to define the metropolitan areas, in agreement with the communes and the interested provinces, and to revise the provincial districts when the metropolitan area does not coincide with the province. A law passed in 1995 gave the regions a deadline of 31 July 1996 for the definition of the boundaries of the nine metropolitan cities. In the event that this deadline is not met, central government will proceed with the definition of the boundaries, which, in case of further delays, are to coincide with the boundaries of the respective province. The governing boards of the metropolitan cities are to be nominated soon after the definition of the metropolitan boundaries.

The functions of programming and town and country planning The task of defining the general objectives of socio-economic and land planning, and drawing up boundaries, are allocated to the regions according to the amount of money made available by local bodies for investment programmes. Communes and provinces work together towards these objectives. The province works with the communes and coordinates their proposals for economic, land and environmental planning of the region. It coordinates the formation of regional plans, formulates its own multi-year programmes both at general and sectoral levels, and it prepares and adopts a *Piano Territoriale di Coordinamento* (PTC), a coordinated land plan that covers all the communes encompassed by the boundaries of the province.

This plan indicates the different proposed uses of the land, the principal local infrastructure and communications, access to underground and surface water supplies, with particular regard to geological stability and drainage. It also is concerned with areas in which there might be opportunities to create parks or nature reserves. The PTC and the multi-year programmes are approved by the region. The province coordinates any communal plans included in its own area and ensures their compatibility.

Mountain communities Mountain communities have statutory autonomy in the field of national and regional laws. They take on plans and proposals for work extending over many years, and operate together to draw up coordinated proposals for land redevelopment. The plans of the mountain communities are approved by the province to which they belong, and the region provides the financing.

Management of services The law allows the possibility of convening meetings between local bodies to carry out services within the commune. At such meetings, in addition to agreements to work with local consortia, action groups can be set up for acquiring public capital.

Programme agreements This deals with another very important innovation. Agreements can be reached for carrying out large-scale works for which more administration has to organized on different levels by either public state businesses or local businesses, or both. Such agreements can be promoted by the president of the region or province or by a syndicate (in relation to their primary or main responsibility) and have the aim of coordinating the work, and determining the timescale, sequence and finances. The agreement is drawn up by a conference of representatives of interested bodies.

Participation and information The region dictates consultation methods, including local advisory referenda when necessary. The citizen's right to information is effectively preserved through the right of veto over all administrative decisions.

Financial autonomy As a matter of principle, communes and provinces have financial autonomy based on certain defined resources, both their own and those allocated to them. The actual structure of finances, based essentially on allocations, will have to be thoroughly revised.

Rights of ownership of land and property

The constitution sanctions the freedom of private enterprise, but states at the same time that this cannot be at the expense of social needs, security, liberty and the dignity of each individual. It allows for decision-making by legal processes as well as direct action in both public and private sectors.

Recognizing its social function, the constitution upholds the right to private property. At the same time, it allows the possibility of compulsory takeover of private property if three conditions are met: the public body acquiring the property must be able to justify its action on the grounds of public interest; the property itself and the takeover process must be legal; and compensation must be paid to the owner.

The guidelines for assessing the amount to be paid in case of compulsory purchase have been the subject of endless legal wrangling. After a ruling by the Constitutional Court in 1980, which declared the previous methods to be against the constitution, new methods have been established. The right to property remains traditional. It includes development rights, despite the fact that, after the issue of the law of 1977, radical legal decisions, as well as the previous unwritten rules, have maintained the distinction between property rights and developmental rights (*jus aedificandi*). This was based on the fact that the new law, albeit burdensome, replaced with the "planning concessions" the planning permissions allowed for under existing laws. These concessions are granted by the commune where the proposed development is located, and are issued subject to the payment of "primary urbanization" charges (towards the cost of water and sewage, roads, etc.), a share of the expenses towards "secondary urbanization" (towards the cost of primary schools, public service buildings, etc.), and a percentage of the building costs. This is interpreted by some as the equivalent of the British example established by law in 1947, whereby the developer has certain responsibilities towards the local authority. On the other hand, according to the Constitutional Court (*Corte Costituzionale*) in order to make such a change in the structure of real estate property rights, there had to be explicit legislation. It could not be based on individual interpretations. Thus, the entire jurisdiction over development rights is the subject of a reform proposal, which is currently being debated.

The present procedure for determining the amount due to be paid in cases of compulsory purchase is described in Chapter 3.

1.2 The economic framework

The fundamental characteristics of the Italian economic system will be presented in this section. It follows the evolution of the 1980s and of the first half of the 1990s. The predominant role of the public budget will be described in order to make its function within the national economy clear.

Economic development data

In the late 1970s, the gross national product grew at a fairly constant rate: from 1976 until 1980, it averaged 4.8%. In the early part of the 1980s, the situation changed to one of virtually no growth. After 1983, the situation improved with modest rises being recorded. 1988 represents the peak year (GDP 4.2%). In 1989 the growth rate slowed down, and this trend was maintained into 1990, when it was estimated at 2%. Imports showed a tendency to decline at the end of 1983, but in 1984 they grew sharply and continued to grow in the following years; in 1989 they reached nearly 29% of the GDP.

The annual growth rate in family spending has not directly mirrored the growth in GDP. However, over the whole period from 1980 family spending has only increased by 2% of GDP. Fixed gross investments showed a decrease during the years 1981–3, and only in 1983 did they reach and then surpass the values of 1980; they continued to grow during the following two years at a fairly steady rate. As a proportion of the GDP, they fell from over 24% in 1980 to around 21–22% during the middle of the decade, reaching 23% only in 1989.

Exports improved in 1984 and 1989. Analysis of the data shows that family spending tended to grow more quickly than exports and investments. Moreover, a causal link between the growth of the latter and the growth in GDP is not supported by the evidence. Exports increased significantly in the period 1993–5, as a consequence of the devaluation of the Italian lire against other European currencies in 1992. In 1995 Italy achieved the highest growth rate of GDP within the EU.

Inflation and interest rates

The late 1970s and the early 1980s in Italy were characterized by abnormally high inflation and interest rates. Even though the trends then decreased until 1987, the levels stayed remarkably high compared to other European countries. In real terms the relative cost of borrowing grew until 1987 and then decreased during 1989–90. Real interest rates appear to have influenced private investment (Table 1.4). Since 1987 the growing need to finance public debt has helped to sustain high interest rates. Savings attracted by the Treasury were obviously not available for investment in the equities or for direct investment in the production of goods and services.

The most striking change is undoubtedly inflation, which reduced from 21.1% per cent in 1980 to 3.9% in 1994. This is quite an achievement for the Italian economy, whose best performance was probably made after the exit from the European Monetary System in September 1992. In fact, in spite of a devaluation against the major European currencies of about 30%, the internal prices of commodities and services remained stable and in some cases even declined in absolute terms. At the beginning of 1995 the rate of inflation set as an annual goal by the government was 5.5%. It was basically achieved. For the second part of the 1990s, the inflation rate is expected to be around 4%.

Family spending

There have been notable changes in consumption patterns since the 1980s. In particular, expenditure on food declined constantly as a percentage of total household consumption. It fell from over 26% of the household budget in 1980 to little more than 20% in 1989, and to about 15% in 1995. However, there are still some important geographic variations, as in the North of Italy the proportion of household budgets spent on food is now below 10%, whereas in the South it is still

Table 1.4 Inflation and interest rates.

	1980	1981	1982	1983	1984	1985	1986	1987	1988	1989	1990	1991	1992	1993	1994
Inflation rate	21.1	18.7	16.4	15.0	10.6	8.6	6.1	4.6	5.0	6.6	6.1	6.4	5.4	4.2	3.9
Rates of interest															
Treasury bonds (med.)	15.92	19.70	19.44	17.89	15.37	13.71	11.40	10.73	11.13	12.58	12.38	12.54	14.32	10.58	9.17
Deposits (max.)	14.86	17.60	18.82	17.62	16.14	14.48	11.97	9.93	9.51	9.32	9.57	9.33	9.99	8.84	6.97
Loans (min.)	19.03	21.44	21.62	19.44	17.64	16.36	14.08	12.06	12.11	12.01	11.78	11.35	12.53	10.48	8.35
Prime rate ABI	19.93	22.13	21.54	19.19	17.67	16.55	14.18	12.74	12.76	13.83	13.35	12.84	14.36	11.40	9.27

Source: Bank of Italy.

Table 1.5 Gross domestic fixed capital formation by product (billions of lire at 1980 prices).

	1980	1981	1982	1983	1984	1985	1986	1987	1988	1989	1990	1991	1992	1993	1994
Construction	49611	48489	45710	46380	45933	45717	46205	45717	46286	47958	50093	50811	49722	46565	44150
Dwellings	25991	25960	24855	25884	25734	24841	24437	23831	23514	23748	25065	25821	25873	25630	25069
Non-residential buildings and civil engineering works	23620	22529	20855	20496	20199	20876	21768	21886	22772	24210	25052	24994	23838	20892	19021
Machinery & equipment	35902	33143	31662	31078	34820	35893	36790	41770	47183	49831	50882	51541	50738	41573	44386
machinery & metal equipment	32183	29626	28428	28124	31960	33179	33808	38467	43319	45622	46948	47740	47017	38463	40998
other products	3719	3517	3234	2954	2860	2714	2982	3303	3864	4209	3934	3801	3721	3110	3388
Transport equipment	8549	9464	8994	8107	8624	9040	9073	9891	10477	11487	12071	11388	11360	8599	8464
motor vehicles	7018	7771	7622	6392	6706	7171	7251	7736	8584	9251	9483	9144	9004	6383	6480
other means of transport	1531	1693	1372	1715	1918	1869	1822	2155	1893	2236	2588	2244	2356	2216	1984
Total	94062	91096	86366	85565	89377	90650	92068	97378	103946	109276	113046	113740	111820	96737	97000

Source: ISTAT.
Note: As of 3 May 1996 the following exchange rates apply: currency unit Italian lire.
ECU 1.00 = 1918.02, DM1.00 = 1019.33, FF1.00 = 301.85, UK £1.00 = 2345.32, US$1.00 = 1567.52.

Table 1.6 Foreign trade balance (as % of GDP at current prices).

	1980	1981	1982	1983	1984	1985	1986	1987	1988	1989	1990	1991	1992	1993	1994
Balance of trade	−3.71	−2.83	−2.19	0.59	−1.41	−1.30	0.78	0.02	−0.09	−0.19	0.11	−0.02	0.26	3.35	3.47
Balance of current payments	−2.18	−2.20	−1.54	0.41	−0.56	−0.94	0.34	−0.29	−0.76	−1.36	−1.55	−2.05	−2.28	1.15	1.52

Source: Bank of Italy.

Table 1.7 Gross added value at market prices by branch (billions of lire at 1980 prices).

	1980	1981	1982	1983	1984	1985	1986	1987	1988	1989	1990	1991	1992	1993	1994
Agriculture	22305	22433	21842	23663	22803	22904	23363	24126	23307	23923	23018	24689	25254	24808	24864
Fuel and power	15053	14963	14636	13864	13900	14082	14676	15082	15402	15720	16381	16598	17167	17238	17704
Manufact. prod.	107810	105941	105161	106001	110747	114046	116656	121162	130213	134093	137511	137211	137487	134194	141308
Construction	28458	28264	27328	26810	25859	25487	25707	26091	26528	27128	28218	28548	28291	26648	25434
Market services	164780	168911	171719	174048	181647	188715	194529	200871	209276	217648	223241	226604	230525	232871	236053
Non-market services	45730	46729	47525	48168	48977	49654	50303	50779	51430	51819	52462	52928	53241	53254	53283
TOTAL	384136	387241	388211	392554	403933	414888	425234	438111	456156	470331	480831	486578	491965	489013	498646

Source: Bank of Italy.

Table 1.8 Working population rate.

	1980	1981	1982	1983	1984	1985	1986	1987	1988	1989	1990	1991	1992	1993	1994
North	43.0	43.4	43.2	43.3	43.7	43.6	44.0	44.5	44.9	44.5	44.4	45.1	45.1	44.1	43.9
Centre	39.7	40.5	40.9	41.3	41.8	42.3	42.6	42.5	43.2	42.9	43.0	43.8	43.6	41.3	41.0
South & Islands	35.9	35.9	36.1	36.7	36.5	37.0	37.7	38.0	38.5	38.5	38.6	38.6	38.7	35.4	35.1
Italy	39.8	40.2	40.3	40.6	40.7	40.9	41.5	41.8	42.0	42.0	42.0	42.4	42.4	40.4	40.1

Source: Bank of Italy.

Table 1.9 Employment by branch (thousands).

	1980	1981	1982	1983	1984	1985	1986	1987	1988	1989	1990	1991	1992	1993
Agriculture	2994	2845	2683	2745	2687	2581	2562	2508	2394	2297	2235	2237	2132	1984
Fuel and power	190	192	190	187	190	193	196	200	199	195	197	196	192	184
Manufacturing prod.	5966	5750	5602	5382	5140	5070	5043	4993	5055	5075	5140	5040	4853	4619
Construction	1762	1792	1791	1779	1675	1652	1633	1615	1604	1593	1634	1681	1700	1665
Market services	7504	7748	8105	8349	8762	9080	9282	9432	9634	9758	9869	10063	10051	9830
Non-market services	3646	3734	3811	3882	3960	4038	4071	4129	4201	4215	4252	4299	4348	4339
Total	22062	22061	22182	22324	22414	22614	22787	22877	23087	23133	23327	23516	23276	22621

Source: ISTAT.

over 20%. Since 1980, there has also been a decline in the proportion of the household budgets spent on clothes, furnishings and fittings, although this is not substantial. In contrast, spending on rent and heating increased. So too did spending on health, entertainment and, to a lesser extent, transport and communications. The percentage reduction in spending on foodstuffs is indicative of a growth in prosperity, and the sustained growth in health spending is probably linked to the inefficiency of the public health service. The growth in spending abroad and the decrease in consumption by non-residents, are linked first to the propensity of Italians to holiday abroad and secondly to the downturn in tourism in Italy in recent years.

Investments

It is interesting to compare the GDP with gross internal industrial investment (see Table 1.5). With the growth of the GDP, the investment sector with the highest growth has tended to be machinery and metal components. Non-residential building and public works, motor vehicles, other manufactured products and transport all showed a less pronounced growth. On the other hand, residential building showed a negative trend; investment decreased during this period, as did capital values.

Foreign exchange

During the 1980s the industrial balance and the total balance of payments both showed consistently negative movements (see Table 1.6). More positive values can be found in the second half of the decade (especially in the industrial balance). The deficit is more evident in the energy sector. In fact, the declared imports of electrical energy grew very rapidly during this decade, from 3.4% of the net energy in 1980 to 14.7% in 1989. The data registered in the 1980s for the balance of trade and the balance of current payments were generally negative. The balance of trade produced a positive result only in 1986 and 1987, whereas the current payments showed a positive balance in the years 1983 and 1986. The trend in the 1990s is quite different, notably after the devaluation of the Italian lira against the other major European currencies at the end of 1992. In 1994 the balance of trade and the balance of current payments reached all time positive values. It appears that the devaluation of 1992 had just an initial influence on the outlook of the Italian economy. Of longer lasting influence was the agreement on labour costs reached between the very influential trade unions (*CGIL*, *CISL*, *UIL*) and Confindustria in July 1993, and by a parallel agreement between the trade unions and the government to restrain within agreed limits the annual price increases within the public utilities.

Economic sectors and value added

Gross value added in Italy originates from the different economic sectors in a way typical of a post-industrial society. The prevalent sector is marketing services (46.3%), with manufacturing industry lagging behind at (28.5%). The relative decrease in the value added of manufacturing is clearly associated with the late 1980s. During the same period the effect of the service sector was more uniform, while the contribution of value added from agriculture and building construction declined steadily.

Using constant prices it can be seen from Table 1.7 that the value added in the agricultural and energy sectors was at the end of the decade little higher than at the beginning, and the construction sector was even lower. Growth in the marketing services sector was the highest, standing at 3.1% per year on average, with manufacturing more limited at 2.5%. The recent structural changes within the Italian economy can be fully appreciated with a comparison of the different weight of the gross added value produced in 1980 and in 1994 by two significant branches of the economy: the manufacturing sector and the market services. In 1980 the observed values were 107810 and 164780 billion lire respectively that is the output of the service sector was 53% higher than the manufacturing sector. In 1994, the gap had increased to 67%.

Employed and unemployed populations and employment by economic sector

To underline the marked interregional divisions which are one of the main characteristics of Italy, comparative data on the employed and unemployed populations are presented separately. Tax revenues from employment are at very different levels throughout the three territorial districts: clearly high in the North, average in the Centre and, most importantly, low in the South and Islands. Still more pronounced are the different unemployment rates. In the North, unemployment follows the fluctuations in the complex economic activity of the country. It has varied from near full employment at the beginning of the 1980s, then going through a period of crisis, until an improvement occurred towards the end of the decade (6% in 1989). In the Centre and South, on the other hand, the number of unemployed grows steadily. It shows no tendency to decrease, even given the more positive economic climate at the end of the decade. The figures of 12% in central Italy and, above all, 21% in the South and Islands clearly demonstrate the precarious industrial base in these localities. The unemployment rate rose from 7.3% to 12.0% between 1980 and 1989, declining to 11.3% in 1994. The novelty of the 1990s is a phenomenon common to most major western countries, that is, the increase of all the major economic indexes (e.g. gross added value and consumer spending) without a substantial fall in unemployment.

The distribution of employment by sectors has changed most clearly in agriculture, where it is steadily decreasing. However, there is still an imbalance between the numbers of employed and their contribution to added value: in 1989, 9.9% of those employed produced an added quota of 5.1%. Variable fluctuations in employment were registered in the energy sector. The reduction of employees in manufacturing was continual (on average for the 1980s -1.8% per year), with a slight upswing in 1989. During that year 21.9% of the manufacturing workforce was producing 28.5% of the added value in the economy. A reduction in permanent employment was seen in the construction industry. Service employment showed a steady growth (on average about 3% per year), to reach 42.2% of the total by 1990. It contributed 46.3% towards added value. Even employment in non-market services grew, but at a slower rate. In 1993 employment in the marketing services branch had reached almost 10 million, increasing from 7.5 million in 1980, while in the same period employment in the manufacturing sector declined from 6.0 million to 4.6 million.

Waged work, added value and wages, timetables and trade disputes

The number of employees fell in all sectors except agriculture. In absolute values, the 713000 self-employed workers (more in 1989 than 1980) were the result of a marked decrease in agricultural employment (-512000) and, to a lesser extent, in industry (-15000). This was more than compensated for by the large growth in self-employed work in services (1241000). The figures for the share of added value per employee allow a couple of comparisons to be made regarding productivity. The highest values can be found in the service and insurance sectors. Here, after years of decline there has been a renewal of growth (10.8% in 1990 compared with 1980).

In the energy sector, productivity being linked to intensive capitalization, also registered fairly high values, although recently was only 4.1% greater than the 1980 figure. In transport and communications, productivity had risen by the end of the decade (17.5%) while the most consistent growth in productivity can be seen in manufacturing (46.8%). The increased efficiency of this sector can also be illustrated by an analysis of the employment trend. As mentioned above there were 6.0 million employees in manufacturing in 1980; the figure was reduced to 5.0 million in 1990 and to 4.6 million in 1993 (Table 1.9). The figures illustrate the fundamental restructuring processes and the technological and organizational changes that have increased productivity during the decade. The various service sector is the sole exception, where the added value per unit of work was less in 1989 than in 1980 (-8.4%). These various services are clearly marginal and almost redundant. The data shows moreover that the growth of employment in the service industry is not linked to efficiency. A similar situation is illustrated by the figures in commercial services, hotels and public works, where productivity, except for the last year, remained lower throughout the decade than in 1980.

The construction industry registered only modest growth in productivity (5.5%), whereas agriculture, still at the bottom of the league, showed a marked growth in added value per unit of work (39.8%). An analysis of productivity and gross wages per unit of work shows that added value grew much more rapidly than wages in agriculture, manufacturing and, to a lesser extent, transport. The wage per work unit increases more than the added value in hotels and cleaning work, energy and the construction industry. During the decade there was a slight reduction in the hours in the working week (by a little more than 3%). In the same period, the number of trade disputes fell.

It should be noted that the total number of hours lost through strikes remained above 60 million per year until 1984, with peaks of well over 100 million hours in 1980 and 1982. Since then, this measure of social tension declined rapidly to an all-time low point of less than 20 million hours in 1992, and about 23 million

Table 1.10 Unemployment rate.

	1980	1981	1982	1983	1984	1985	1986	1987	1988	1989	1990	1991	1992	1993	1994
North	5.1	6.1	6.7	7.7	8.1	8.0	8.0	7.8	6.9	6.0	5.1	5.1	5.9	6.2	6.8
Centre	7.5	8.3	8.6	9.1	9.0	9.1	9.7	9.7	9.8	10.6	9.8	9.6	9.9	8.6	9.6
South & Islands	11.5	12.3	13.0	13.8	13.6	14.4	16.5	19.2	20.6	21.1	19.7	19.9	20.4	17.5	19.2
Italy	7.6	8.5	9.1	9.9	10.0	10.3	11.1	12.0	12.0	12.0	11.0	10.9	11.5	10.2	11.3

Source: Bank of Italy.

Table 1.11 Annual contractual work time in industry (hours per employed).

1980	1981	1982	1983	1984	1985	1986	1987	1988	1989	1990	1991	1992	1993	1994
1792.7	1783.4	1778.1	1778.0	1773.6	1753.5	1750.6	1749.5	1748.2	1739.3	1735.4	1735.4	1734.6	1733.6	1727.4

Source: Bank of Italy.

Table 1.12 Strikes (thousands of hours).

1980	1981	1982	1983	1984	1985	1986	1987	1988	1989	1990	1991	1992	1993	1994
115201	73691	129940	98021	60923	26815	39506	32240	23206	31053	36269	20895	19510	23880	23618

Source: Bank of Italy.

Table 1.13 Public debt (billions of lire at current prices).

	1980	1981	1982	1983	1984	1985	1986	1987	1988	1989	1990	1991	1992	1993	1994
Absolute values	228240	283130	361466	455361	561489	683044	793583	910542	1035811	1170134	1318811	1487476	1676150	1866272	2046510
As % of GDP	58.46	60.49	66.31	71.89	77.37	84.27	88.19	92.55	94.87	98.05	100.51	104.06	111.45	120.39	124.70

Source: Bank of Italy.

in 1993 and 1994. In spite of some common beliefs, which probably date back to the 1970s, strikes are less and less common in Italy, being replaced whenever possible by a continuous dialogue between trade unions, employers and central government. It is a fact that there have been fewer strikes in Italy in the first half of the 1990s than in France and Germany).

The public deficit and the consolidated budget of public administration

The growth of the public deficit is obviously very serious for the Italian economy. Its tendency to outstrip the GDP appeared in the 1980s and was not stopped until 1995, when the government led by prime minister Lamberto Dini managed to achieve a reversal in the historical tendency of the public deficit to grow. Between 1980 and 1994 the public deficit grew from 58.5% of GDP to 124.7% of GDP. The worst years appear to be the 1980s, with governments led by the Christian Democrats and by the Socialist Party. It has been suggested, quite convincingly, that the increase in public spending grew at the rate as the decline in the support for the ruling parties. It must be remembered that a significant share of public spending went to finance marginal activities and well-targeted social groups in a silent exchange of voting support. For example, thousands of state pensions have been given for years to people falsely claiming to be handicapped, at an estimated cost of 20–30 thousand billion lire per year. The collapse of the party system between 1992 and 1994 was a necessary condition for the reversal in the public spending trend. In the past decade, the inevitable drain on resources through the financing of this deficit has had some negative effects. On the one hand, the alignment of financial shares to inflation has been fairly limited. This has produced "reversed redistribution" effects, which have struck the incomes of medium- and low-paid employees. On the other hand, the need to keep interest rates high has altered the amount of savings invested by industry.

The figures for the consolidated balance of public administration show that current spending tends to grow more quickly than current returns, despite a consistent growth in fiscal returns, especially direct returns (Table 1.12). The categories of spending that grow most rapidly with the growth of GDP are as follows: social services, salaries and incomes, private consumption and interest paid on state bonds issued to cover the public deficit. If spending is calculated on a percentage basis, it can be shown that for social services and intermediary consumption it remained fairly constant (the first around 33–34%, the second 9–10% of total spending). Salaries and incomes tended to fall from little more than 26% just over 23%. Spending on interest payments rose from 12.5% in 1980 to 17.2% in 1989. Despite the good intentions expressed every year by the Ministry of Finance, the vicious circle produced by mass accumulated debt was fairly difficult to break, until the mid-1990s, when the combined effect of curbed interests rates and a lower public spending level led to a moderate decline in the global amount paid annually by the state for debt servicing.

Investment in the wider public sector

Investment in the wider public sector (WPS) shows a figure higher than all other investments. It rose from 17% to almost 25% during the decade. The most consistent investment figures are those of the communes and the provinces (in 1989 they made up almost 30% of the WPS). It can be shown that the investments of the communes and provinces show the most marked tendencies for growth, followed closely by public production companies (including the state and the central regions) which all show a fairly similar and steady trend. This seems to indicate a tendency towards fragmentation in the investment policies of public bodies and it appears to contradict their responsibilities. Traditionally, the communes are administered as "providers of public services", but in fact they make up part of the wider investment. The regions meanwhile, initially planning bodies, have a fairly modest role to play in this field.

Short-term forecasts The major short-term forecasts presented at the beginning of the 1990s are given below (Table 1.14). They could not take the Gulf crisis into account, and so they proved to be rather optimistic. The first half of the 1990s was for the Italian economy a period of recession, a contributory element being the slowing down of public spending. This started in 1992 with the uncovering of a bribery system in the public works sector (*Tangentopoli*) and it was followed by similar scandals in several other areas of the economy.

Table 1.14 Comparison of forecasts of key economic rates for 1991 and 1992.

	Government		ISCO		Confindustria		Prometia		OECD	
	1991	1992	1991	1992	1991	1992	1991	1992	1991	1992
GDP	2.7	3.0	2.5	2.0	1.4	1.7	1.6	2.9	2.4	2.7
Internal demand	2.7	3.0	n.a.	n.a.	2.2	2.4	2.0	3.3	2.5	2.8
Private consumption	2.7	2.9	2.7	2.5	2.4	2.6	2.7	3.0	2.6	3.0
Investments*	4.4	4.7	4.0	3.5	2.0	2.3	1.5	6.2	4.4	4.2
Export	5.9	6.5	5.0	4.5	2.7	3.6	4.1	5.9	5.2	6.3
Import	6.0	6.2	5.5	5.0	5.4	5.5	5.0	6.8	5.3	5.8
Inflation	5.0	4.5	6.2	6.7	6.5	6.6	6.6	6.4	6.8	5.8

* Investments are net of the building sector for *ISCO*, *PROMETIA* and OECD.

Table 1.15 reports the major forecasts for the Italian economy for the period from 1995 onwards. There is a consistent view that GDP growth will be around 3% in 1995, and for 1996 there is a range from 2.6% (*ISCO*) to 3.4% (European Committee). For 1997 and 1998 the research institute Prometia expects growth to be between 2.2% and 2.8%. The outlook for the Italian economy appears good for the late 1990s. This is encouraging, given that all the forecasts take into account the fact that the devaluation of the lira, which occurred in 1992, has already ceased to affect the internal economy. This is well demonstrated by the forecast for the growth of exports for 1995 (generally predicted to exceed 10%) and for subsequent years (6–8%). This positive outlook is even more optimistic if it is recognized that it takes place, despite the rise in imported energy costs, which are very significant for the Italian economy.

Table 1.15 Italian economy, main forecasts as of July 1995.

	1995	1996	1997	1998
GDP *(deflated values)*				
CER	2.9	2.6	3.1	
Prometia	3.3	2.6	2.2	2.8
European Committee	3.3	3.4		
IMF	3.1	3.1		
IRS	3.0			
ISCO	3.1	2.6		
Wefa Group	3.1			
CSC Confindustria	3.2			
DRI McGraw-Hill	2.8			
JP Morgan	3.0	3.0		
Household consumption				
cer	1.9	2.0	2.7	
Prometia	1.9	2.1	1.5	1.9
IMF	2.6	3.0		
IRS	1.9			
ISCO	1.9	1.9		
Wefa Group	2.0			
CSC Confindustria	1.7			
DRI McGraw-Hill	0.9			
JP Morgan	1.6	2.2		
Gross fixed investments				
CER	6.3	4.6	5.4	
Prometia	4.5	5.4	3.4	3.6
IMF	3.1	3.1		
IRS	4.1			
ISCO	4.6	5.0		
Wefa Group	5.5			
CSC Confindustria	6.9			
DRI McGraw-Hill	4.6			
JP Morgan	4.4	5.5		
Exports				
CER	8.0	6.6	6.1	
Prometia	10.4	7.5	7.3	6.9
IMF	10.0	6.5		
IRS	10.5			
ISCO	9.7	7.7		
Wefa Group	8.0			
CSC Confindustria	10.4			
DRI McGraw-Hill	8.8			
JP Morgan	9.3	8.1		
CPI				
CER	4.8	4.6	4.5	
Prometia	5.2	4.6	3.8	3.4
European Comittee	5.2	4.5		
IMF	5.8	3.3		
IRS	5.2			
ISCO	4.7	4.2		
Wefa Group	5.0			
CSC Confindustria	5.7			
DRI McGraw-Hill	5.2			
JP Morgan	5.5	6.0		

Table 1.15 (continued).

	1995	1996	1997	1998
Imports				
CER	7.5	5.9	5.9	
Prometia	8.5	8.1	5.5	5.7
IMF	8.5	6.5		
IRS	8.6			
ISCO	8.1	7.1		
Wefa Group	8.0			
CSC Confindustria	9.1			
DRI McGraw-Hill	6.8			
JP Morgan	8.2	6.8		
Unemployment (rate)				
CER	11.4			
Prometia	12.0	11.6	11.1	10.5
European Comittee	11.4	10.9		
IMF	11.3	10.6		
IRS	11.3			
ISCO	11.4	11.2		
Wefa Group	11.0			
CSC Confindustria	11.1			
DRI McGraw-Hill	11.9			
JP Morgan	14.4	13.8		

The growth of GDP and exports is expected to occur without creating further inflation; the rate is expected to decline gradually from 5% in 1995 to 3% in 1998, thus bringing Italy closer to other European countries. At the end of 1995, although the official statistics are not yet available, it seems that the forecasts of the *CPI* will prove to be basically correct. The same sources are much more conservative in predicting the future of household consumption, which is expected to increase by roughly 2% per year. One area where the figures are unanimously negative concerns unemployment, which is expected to remain well above 10% until 1998. In this domain, it seems that the growth of production does not mean a parallel rise in employment. This phenomenon is not uniquely Italian, as it started to occur in the USA in the late 1980s, and in the major European economies at the beginning of the 1990s.

Trends of reforms of local finance

From the mid-1990s there will be a trend towards local government gaining increased taxation powers. One of the instruments to achieve a more balanced role in the taxation of properties, personnel incomes and corporate profit is the *Imposta comunale sugli immobili* (*ICI*), the local tax on real estate properties. Since 1993 this has acted as a replacement for the *Imposta straordinaria sugli immobili (ISI)*, the real estate property tax. *ICI* is imposed on all real estates in a given commune, with some exemptions (churches, properties of non-profit institutions, etc.). Each commune is free to fix the level of the *ICI* in a range 0.4–0.7% of the cadastral value of the property. In the near future it is likely that this range will be revised upwards and that the communes will be able to differentiate the tax rate between different parts of their territory. The tax is levied directly by the communes.

Nevertheless, the increasing responsibilities of local government seems to exceed the likely available budget, even when this is enlarged by the new *ICI* local tax. In spite of some common beliefs, the fiscal pressure in Italy (measured in share of GNP taken by government) is one of the highest in Europe and it is unlikely that it could be increased significantly. Thus, there are various proposals to allow the local government to issue public bonds in order to raise additional revenues.

This is not new. In 1881, the city of Naples issued a 99-year bond (at a fixed rate of 5%), to finance public works in the area of sanitation and urban renewal. This practice has been repeated until recently by major cities such as Milan, Genoa and Naples. For example, the city of Milan issued a 20-year municipal bond at a fixed rate of 7% for the amount of 100 billion lire in 1975. Towards the end of 1970s, however, local government ceased to operate in the capital markets, because of tighter controls by the central bank and the centralized taxation system.

The municipal bond is the second market in the USA, just after the treasury bonds. In Europe, cities in various countries have recently issued municipal bonds and the major Italian cities look set to follow. The financial law of 1995 reintroduced the possibility, for the communes, to issue *buoni ordinari comunali (BOC),* ordinary municipal bonds. The new rules, specified at the end of 1995, will differentiate the municipal bonds from the treasury bonds; mainly in the reimbursement procedure, where the capital part will not be paid back in one final instalment (as is the case of treasury bonds) but in periodical instalments (every three, six or twelve months), together with the interest. It is expected that the major cities will voluntarily ask for a financial rating, especially given their desire to open up their municipal bonds to the Eurobond market.

Nevertheless, it would be a mistake to overemphasize the effect of a classification by rates of the Italian municipal bonds, because the overwhelming share of the bond market is made up of Italian treasury bonds for which the risk of default is remote, if existent at all. In fact, Italian public debt has some specific characteristics that are usually overlooked:

- It is almost entirely in the portfolio of Italian households, whereas most of the treasury bonds of countries usually considered better off than Italy in terms of public debt (e.g. the USA, Japan, Belgium) are owned by foreign investors.
- The Italian households (i.e. the owners of the state deficit) own assets worth twice the value of the public debt, whereas households in the USA, Japan and several European countries own a much smaller share of assets.
- The estimates of the public deficit do not take into account the value of the real estates owned by the state and which could be put on the market, because they are not essential for the functioning of the state or because they are completely unused. This wealth is roughly estimated to be worth half a million of billions lire, and this figure does not include the real estates owned by the state but unavailable for sale because they are strictly necessary for public services. In January 1993, the government set up *Immobiliare Italia,*

a company whose mandate is to sell in the open market part of the state-owned real estate. To date, the company has not yet sold a single property, as the Ministry of Finance has not yet allowed it to sign the required general agreements with various institutions. The privatization of the economy is, however, a constant option of all recent governments. The giant insurance company *INA*, for example, was partially sold to private stockholders in 1994, and the same has happened or will happen to other major companies owned by the state: *COMIT* (banking), *STET* (telecommunications), *ENEL* (electricity), *ENI* (oil), to name a few, plus the remaining shares of *INA*. This trend will, sooner or later, show its practical effects in the real estate sector, although selling shares on the Stock Exchange is much easier than selling unused properties in the real estate market.

One major difficulty, which will have to be overcome if the privatization is to take place, is a requirement for co-operation between central and local government. The value of many state-owned properties is strictly linked to the use allowed in the area, and this is the responsibility of local government (mainly communes). For example, in many Italian cities there are military grounds and buildings in prime locations which could be put on sale, but no one would buy them unless the local government were willing to grant planning permission for the redevelopment of the area. A recent survey of the real estates owned by provinces and communes has disclosed that the properties of these categories of local government have very limited rental returns: 3.3% per year for the provinces and 1.4% per year for the communes, but in terms of actual money flows, the returns drop to 0.3% for the provinces and 0.8% for the communes, because of enduring rents arrears, squatters and freeholdings granted for social reasons.

One would expect that the same households who own the Italian treasury bonds will also be the major buyers of Italian municipal bonds. In current circumstances it is likely that the households will act according to their perceptions of financial convenience associated with low risk bonds, rather than to "ratings" issued by a distant agency according to standard mathematical calculations. The image of cities, and the associated "fame" of good government practices, will play a major role in moving financial flows from one municipal bond to another. Also, the existence of a real secondary market, where it is possible to cash municipal bonds before their expiry date, will be a major determinant of the success of the Italian municipal bond market.

1.3 The social framework

Demographic development

At the time of national unification in 1861, the population of Italy was slightly more than 25 million. By the beginning of 1987 it had reached 57291000, thus placing Italy in second place after Germany among the most populated west European countries. In more recent periods, however, a substantial decrease in annual population growth rates has been recorded. The average annual population increase in the period 1981–7 was only 2.5 inhabitants per thousand, whereas in the years 1971–81 it was 4.4 per thousand, and for 1961–71 it was 6.7 per thousand. The growth process has declined largely because of significantly lower birth rates and a constant reduction of mortality rates (at least since the end of the Second World War). As Table 1.16 shows, this has caused a decrease in the yearly positive balance of births over deaths, from over 400000 to close to zero. In contrast, the births that do occur are the safest in the world. Recent statistics of the WHO and of the UN show that in Italy the infant mortality rate is the lowest in the world: 1:17361 compared to 1:15432 in Norway, the second lowest, and 1:8772 in Australia, the third lowest. Third World countries are lowest in this ranking with Mali, for example, at 1:7. As of 31 December 1994, the Italian population stood at 57266137, of which 63.6% lived in the northern and central regions and 36.4% in the South and the islands.

Table 1.16 Live-born and dead per 1000 inhabitants in Italy 1951–93.

Years	Birth rates	Death rates	Balance live-born/dead
1951–60	17.80	9.60	402631
1961	18.40	9.30	461202
1971	16.80	9.70	383528
1981	11.00	9.60	79601
1991	9.71	9.55	9044
1993	9.68	9.72	–2456

Source: ISTAT.

When we look at the natural changes in the Italian population, we must consider the fact that values calculated on a national level obscure the difference between the northern and central regions and the South. For example, the birth rates in northern and central regions were as high as 7.9 per thousand inhabitants (minimum values of 6.1 were recorded in Liguria, whose regional capital is Genoa), whereas corresponding values in southern regions were 12.8 (maximum values of 14.1 were recorded in Campania, whose regional capital is Naples). In the same year, a natural negative balance was recorded in the North and Centre of Italy (-85535), whereas in the South the balance was positive (96253).

However, these geographical variations should not divert the attention of the reader from the main recent phenomenon: Italy, the country of large families shown in so many American movies of the postwar period, has become over the years the country with the lowest natural growth rate amongst all the industrialized nations. Indeed, the birth rate dropped steadily from around 18.0 per thousand inhabitants in the 1951-61 period, to a mere 9.68 in 1993 (Table 1.16). As in the same year the death-rate was 9.72 per thousand inhabitants, there was in the country a negative balance live-born/dead. It seems that this phenomenon is the outcome of a number of factors which have been caused by some wrong social policies, or, most likely, by the absence of policies in key social domains.

Therefore, it is likely that the Italian households would respond accordingly to a renewed attention to the family. New measures are being proposed to encourage procreation: reduction of the income tax burden on households with two or more children; incentives to find suitable housing; provision of more social services; extension of flexible-time jobs for women. As in the urban planning field, the success of the above mentioned measures will not depend so much on the accuracy with which they are initially designed, but on their actual implementation and continuous monitoring and adjustment.

Mobility

The above changes in the natural balance took place in a phase during which another component of the demographic process, the net migration rate, was going in the opposite direction. Historically, natural population growth in Italy has been balanced by emigration to other parts of Europe and the rest of the world. In 1913, for example, emigration reached a record high of 900000, and in more recent periods it reached a peak of 327000 per year in the period 1957–61. The average number of emigrants has been around 87000 people per year during the 1980s, but this flow has been counterbalanced by a slightly larger immigration factor (92600 people per year from 1971 to 1981, and more in the 1980s) which has resulted in a positive migration balance. In 1994, there was a net migration balance of 44537 people, as 61435 residents left the country, and 105972 were new arrivals.

When considering the most recent migration patterns, the steady growth of foreign immigrants has to be taken into account. Immigration from developing countries is particularly significant. This still finds relatively little place in official statistics, because it is mainly illegal. However, according to *ISTAT*, foreign citizens who in 1971 were only 2.2 per thousand of the total population (121116 people), in 1981 reached 3.7 per thousand in 1981 and 5.7 per thousand in 1986. According to the statistics of the Department of the Interior (*Ministero degli Interni*), residence permits granted by the same Department steadily increased as well: they were 186413 in 1975, 298749 in 1980, 403923 in 1984 and 572103 in 1987. Taking the number of permits granted in 1980 to be 100, the index rises to 120 in 1982, 135 in 1984 and climbs to more than 191 the following year. In the 1988–91 period, the phenomenon has become even more noticeable, and it reached alarming dimensions in 1992–5, when the influx from North African and Asian countries was coupled by the illegal entry of tens of thousands of people from eastern Europe (mostly from Albania and former Yugoslavia). In 1993, legal immigrants reached a total of 987000, or 1.7% of the total population; illegal immigrants are thought to be at least half as many again as the legal ones.

The new immigrants have difficulties in integrating with the native Italians. In 1989 a survey of 500 immigrants who asked for help at *Caritas* in Rome showed that the share of those living with other immigrants increases over time: it was 27.5% for those in Rome for one month or less, and 56.0% for those in the country for two years or more. This trend brings about a worsening of living conditions. About 55% of immigrants interviewed declared that they had the use of no more than 9m^2 of floor-space. At the national level, the immigrants tend to cluster by nationalities of origin in derelict housing in the historical centres. This has prompted conflicts with home-owners and retail and business owners in nearby areas, who fear for the change of image of the neighbourhood and for the consequent decline of property values. Municipal authorities have very limited means to manage the social and economic consequences of migration waves which are completely out of their control.

Most other European countries have a foreign population larger than that of Italy (1.7% in 1993), for example, 8.5% in Germany, 6.3% in France, 3.5% in Great Britain, 5.1% in the Netherlands and 5.8% in Sweden. But, since Italy has been for the best part of this century a country of emigrants, mainly towards other European countries, America and Australia, its political class has not had the opportunity to become psychologically prepared to cope with the opposite phenomenon.

At the end of the 1980s, the so-called Martelli Law *(Legge Martelli)* was passed, and soon proved to be totally inadequate. The law grossly underestimated the actual flows of immigrants and it did not provide the immigration authorities with either the legislative or the material means to regulate and control the growing tide of immigrants. The identity of immigrants from Third World countries or from the collapsed communist regimes of the eastern Europe is not ascertained at the point of entry into Italy through a secure system. As a result, immigrants who commit minor crimes often declare false identities and false nationalities, pretending it is the first time that they have committed any crime in the country, and in the light of the present legislation they are soon released. The phenomenon has reached the point where even more serious offences, such as drug smuggling and control of the prostitution, are rapidly coming under the control of legal and illegal immigrants from poor countries.

The lack of a serious immigration policy is reflected in the fact that, according to recent estimates, over 15%of the population of Italian prisons consists of foreigners. The burden of the immigration phenomenon is carried by the municipal authorities, who, since the early 1990s, have brought pressure on central government for tighter controls at the frontiers and for more revenues to assist the social problems of those already inside the country. Practically all parties have proposed new legislation to address this. Reform of existing legislation is expected soon, in response also to the EU, which has maintained that the controls at the southern flank of its boundaries should be tightened up if the free flow of citizens within the Union is to become a fact within the deadlines set with the Schengen Treaty.

By contrast, internal movements are losing momentum. The 1980s did not see anything like the great migrations from the South to the North so characteristic of the years from 1958 to the early 1970s. Annual interregional movements reached the mean value of 231000 units in the 1959–62 period, subsequently falling to 108000 in the 1971–4 period and to 43000 in the 1979–82 period, stabilizing at around 30000 units in 1984. More specifically, industrial regions of the Northwest, which were accommodating around 180000 new residents per year between 1959 and 1962, showed a reverse trend with a balance of -13500 in the 1983–4 period. At the same time the negative migratory balance of the South thinned out to -122300 in 1959–62 and -13400 in 1983–4. Central regions kept their trend towards a positive balance from the 1950s onwards, and northeastern regions went from negative to positive balances in the mid-1960s, with a noticeable reversion in the late 1970s.

With the end of the great South–North influx, overall mobility decreased as well. Changes of residence inside the country, which had reached a maximum of 34.3 per thousand of total residents in the 1959–62 period, have stabilized at around 21–22 per thousand since the mid-1970s. Furthermore, these figures increasingly reflect not only the population shifts between regions, but movement within the great metropolitan areas, where the main factor is the population shift between the inner-city and peripheral areas and nearby towns.

On the whole, from a social and demographic viewpoint, the features that hitherto made Italy abnormal among western European countries are receding. In 1985, for example, the country's natural balance (0.6 per thousand) was roughly half way between the negative balance of West Germany and the positive one of France and the Netherlands. The average life expectancy (70.6 years for men, 77.2 for women) is similar to that of the most developed countries in Europe.

From a social and cultural viewpoint, currently observable trends also put Italy closer to the most developed Western countries. Particularly interesting are the yearly reports in which *CENSIS* (the leading social research institute) reviews current trends inside Italian society, and the work of some sociologists on the behaviour and value-systems of youngsters. From a complex and contrasting picture, the fact emerges that far less importance is now given to work, especially among the young, and that politics and public commitment play a diminishing role. Conversely, the importance given to self-fulfilment in a variety of ways both public and private is growing. These include friendships, use of spare time, relationships with nature, and occasional voluntary work to solve tangible problems such as social marginalization, drugs and the integration of foreign immigrants into Italian society.

1.4 Land and property markets

Owner-occupation

In Italy, owner occupation increased, according to the recurrent ten-year census, from 40% of the total in 1951 to 50.8% in 1971, 59.0% in 1981, 68.0% in 1991 and 71.6% in 1993. Conversely, households in rented homes decreased from 48.7% of the total in 1951 to 44.2% in 1971, falling to 35.5% in 1981 and to 25.3% in 1991 (Table 1.17). The main shift from rented housing to home ownership took place in the period 1971–91. The census also accounts for the houses neither owned nor rented, but used for "other title", that is freehold or occupancy in exchange for services.This is consistent with field surveys made by leading real estate agents like Gabetti, who maintain that nowadays almost nine out of ten home buyers are already homeowners in search of one or two more rooms or higher quality housing or different locations.

Ownership is less common in the urban areas (47.4%) than in the small and medium centres, and in the North than in the South. These trends continue, but in the first half of the 1990s the trend is becoming less marked as the years pass, in spite of a general decline of housing prices, according to the available data. This apparently irrational behaviour can be explained by the fact that those who could afford to buy a house were already home-owners. The motivations to purchase a different house were the search for higher-quality accommodation and for more space. Those who became home-owners for the first time were primarily young couples, evicted households and old people looking for a place where they could invest their savings. It seems plausible to argue that the trend of growing home-ownership is about to level off, because there is in the country a share of population who are not in the position to buy a house: those below the official poverty line (currently 13 million lire per year per person). These amounted to 5.1% in 1970, 8.1% in 1980, reached a peak of 12.8% in 1988 and then declined steadily to 7.8% in 1993. In addition, some households who still live in rent-controlled houses do not have any economic incentive to change their status. Finally, a proportion of the upper income households traditionally find more advantages in living in high-quality rented houses than in owned properties.

The increase in home-ownership testifies to the economic growth of the country, but brings with it some negative consequences. In the first place the growth of ownership, given the current legal and fiscal framework, hinders residential mobility just when labour mobility is becoming more important. The freezing of residential mobility also threatens those "moves" inside the market that allow for a better utilization of one's assets.

Furthermore, the rise of ownership as an indicator of a generalized improvement in housing standards must be viewed with caution.

Certainly, the owner-occupier section of the housing stock seems to be in less need of repair and it is also less crowded, which implies that a shift to ownership brings about an improvement of housing standards. On the other hand, soaring real estate prices and a rigid rental market prevent a portion of low- and lower-middle income families from improving their situation. A growing polarization is thus developing, in urban areas more than anywhere else, between the existing home-owners and those who live in rented houses and who find increasingly difficult to buy their first house. For the most part, the home-owners enjoy good and quite easily improvable conditions, whereas the home renters suffer from poor conditions and have few opportunities for improvement. The available data on living conditions by housing title shows in all cities a lower crowding index in the owner occupied houses compared to rented units (Table 1.18),

Table 1.17 Occupied housing units by title.

	Owner-occupied		Other title		Owner-occupied and other title		Rented housing		Total	
	n ×1000	%	n ×1000	%	n ×1000	%	n ×1000	%	n ×1000	%
Census 1951	4301	40.0	1214	11.3	5515	51.3	5241	48.7	10756	100.0
Census 1961	5972	45.8	984	7.6	6956	53.4	6076	46.6	13032	100.0
Census 1971	7766	50.8	766	5.0	8532	55.8	6769	44.2	15301	100.0
Census 1981	10350	59.0	976	5.5	11326	64.5	6221	35.5	17545	100.0
Census 1991	13419	68.0	1317	6.7	14736	74.7	5000	25.3	19736	100.0
Survey 1992					15740	75.4	5128	24.6	20868	100.0
Survey 1993					15659	75.5	5082	24.5	20741	100.0

Source: ISTAT.

Table 1.18 Living conditions by housing title.

	Owner-occupied	Rented
Turin	0.74	0.91
Milan	0.76	0.92
Genoa	0.59	0.66
Venice	0.69	0.74
Bologna	0.68	0.76
Florence	0.65	0.72
Ancona	0.69	0.74
Rome	0.73	0.86
Pescara-Chieti	0.71	0.82
Naples	1.04	1.14
Bari	0.84	1.02
Taranto	0.86	1.05
Reggio Calabria	0.80	0.95
Catania	0.79	1.00
Palermo	0.76	0.95
Cagliari	0.76	0.91
Total	0.76	0.90

Source: ISTAT data as processed by CRESME.
Note: (1) Owner-occupied includes "other title".

Level and trends of prices

The functioning of rent and ownership markets With the above in mind, it is now possible to describe the complex links between the rental and the ownership markets in Italy. Several direct surveys and data analysis confirm the limited flexibility at the lower end of the house price range, manifesting a sort of solvency threshold to the ownership market. The very inflexibility of these "access prices" is as attributable to an unsatisfied demand of houses to rent expressed by the lower income groups. Given the scarcity of available houses to rent and the rent levels applied, the low-income groups put pressure on the access levels of the ownership market. The link between the rental and the ownership market is thus substantial although indirect: the peculiarities of houses for rent influence the formation of house sale prices, not the reverse. Two propositions are thus put forward. The first one concerns the filtering processes: in such a property market, "filtering up" is only effective in rising housing standards for those who are already owners. In other words, the creation of new housing at the top end of the market may well "free" housing at the lower levels, but will not reduce its relative price. There is thus a structural hindrance to the formation of a "society of owners" that does not depend upon the number of available houses as much as it does on the "access price". The second proposition concerns the relationship between the quality and features of a unit and its price: there is no linear and constant relationship, as trends in different price brackets are different. It follows that a "scientific" determination of the rent as a percentage of the market value of the unit, derived once and for all from its functional standards and location (which is implicit in Italian rent control measures), is too abstract and incompatible with market behaviour. Immediately after the approval of the Fair Rent Act (1978), the rental market was frozen: those who were already inside it benefited from the law; the others had almost no choice other than to become home-owners. This explains why, in the years after 1978, households searching for a roof over their heads bought houses of poor quality and in poor locations. The result was rising prices, even in this sector of the market.

Both information on the rental market and explanatory analysis of its mechanism are scarce, notwithstanding that in Italy the rental market has been the object of many legal Acts. During the discussion and initial implementation phase of the rent control legislation at the end of the 1970s an intense debate on the forms of regulation flourished. However, the problem was not always considered in sufficiently general terms. A very interesting point was made then and, even if it cannot be supported by systematic analysis, it still seems applicable today: the range of family incomes is much narrower than the range of available rents.

A wide variety of functional and locational features is not matched by a corresponding variety of rent options, and the maximum and minimum rent levels do not meet the spending capacity of users. In substandard housing, where the initial investment has been largely paid off, rents appear to be particularly high: they thus become a sort of regressive taxation of incomes. Poor conditions, together with (relatively) high rents, force mid-low-income families to save, abnormally compressing their consumption levels, in order to access the ownership market. And this explains the small flexibility at the lower end of house prices. A very general conclusion could thus be that the first goal for a coherent housing policy should be a wider range of rents at the bottom end. From this point of view, the strict regulation of the rental market has paradoxically done the opposite, flattening the range of rents. Some kind of regulation of the market was probably unavoidable because of the need to overcome a permanent rent freeze. However, the law has ended up in imposing an intrinsically weak abstract theoretical model on the market and, most of all, it has been unable to mitigate the negative mechanisms at the lower rent levels.

The complexity of the rental market is, anyway, also attributable to the need to consider general variables (not all inside the real estate sector), besides the particular relationships of the property market. Houses for rent are considered as part of the vast field of private investment where they compete with other alternatives. Macroeconomic variables too, like inflation rates, trends on financial markets and income distribution, influence this wider investment sector and therefore influence the number of available houses and the rent levels.

Requirements for land

Land consumption for housing in Italy is caused primarily by the expansion of the housing stock, by 2.6 million units in the 1951–61 period, 3.2 million between 1961 and 1971, and by 4.4 million in the 1971–81 period (Table 1.19). Between 1982 and 1987 there were an estimated 220000 new houses per year. In the 1961–71 period, larger increases are recorded for the Centre (30.7%) and in the Northwest (25.88%), while the national average rate was 22.66%. In the following decade (1971–81), expansion was faster in the Islands (36.25%), in the South (27.47%) and in the Northeast (26.85%), with a national average of 25.30%. At national level, the housing stock increased from 14.2 million housing units in 1961 to 24.8 million units in 1991 (Table 1.19).

Examination of the data, disaggregated by spatial location and type of dwelling, for the 1961–71 period shows that the production of new housing was concentrated in the "rings" of metropolitan areas, where the rate of increase was very high (60.59%), whereas in the following decade more moderate increases were again found in the metropolitan rings (31.24%), and also in small towns (31.21%) (Table 1.19). The waves of internal migration of the 1960s (from the South to the North, from the country to the city) explain high levels of construction in the metropolitan rings, the residential explosion of small towns was the new factor of the 1970s and 1980s. In the Centre–North, this widespread growth is the product of two components: first, decentralizing economic activity attracted new residential developments; and, secondly, the expansion of touristic communes meant a sharp increase in holiday and weekend homes. In the South, although the leisure factor still applies, another element arises which is connected to the very poor housing standards in the small towns in the early 1970s: the general rise in incomes has led many families to abandon the old centres for a new home, often built with a less than complete compliance with existing planning regulations. In all three decades observed, the growth that occurred in the category "small cities" (as defined in Table 1.19) is higher than the growth of metropolitan areas. The gap is larger in the decade 1971–81 (21.44% compared to 7.65%) and 1981–91 (10.25% compared to 4.71%) than in the decade 1961–71, when there was a growth of 34.91% for the small cities compared to 30.14% for the metropolitan areas. However, it must be noted that, starting in 1971, the fastest pace of housing development has been constantly observed in the smaller communes, which are indicated globally in Table 1.19 as "Rest of Italy". Here, there was an increase in housing units of 29.84% in 1971–81 and 15.77% in 1981–91. In absolute terms, there was an increase of 6.5 million housing units (from 13.0 million units to 19.5 million units) in just 20 years.

The production of leisure houses in the 1970s was intense (104%),the largest increases being in tourist (152%) and coastal (157%) settlements (Table 1.20). In the 1980s, although there was a decrease in the number of houses built every year (Table 1.21), an increase in the average size of each house was also recorded: from 3.99 rooms (87.3 m^2) in 1984 to 4.19 rooms (92.6 m^2) in 1988 (Table 1.21). On average, the number of units per building declined throughout the period 1980–93, as lower density forms of building were preferred (Table 1.23). For example, the percentage share of housing projects having more than 15 lodgings per building decreased from 6.29% of the total in 1980 to 4.17% in 1993. Also, housing projects comprising just one housing unit decreased from 50.8% in 1980 to 49.8 in 1993, whereas constructions having two units increased from 18.8% in 1980 to 20.0% of the total in 1993, and projects comprising three to fifteen housing units also increased, from 24.1% in 1980 to 26.0% in 1993.

The leading demand in the real estate market of the 1980s has very definite features: it is a demand for substitution for self-use, and for improvement of housing and environmental standards. The most active segments of the market are thus the upper and upper-middle ones, central and semi-central quality homes, and large, one- or two-family units with a private garden in the suburbs. This is an "affluent" kind of demand, one in search of a higher standard while already enjoying a good one. At the other extreme, in the lowest segments of the market the demand is from mid- and low-income households who are entering the property market for the first time.

1.5 Trends of spatial development

National trends

The Italian urban network Central and northern Italy have a density of cities among the highest in Europe, but a real national urban network did not emerge until the 1950s, when interchanges at the national level overcame local and regional networks. The process was spatially uneven, with, by the late 1960s, almost a third of the population concentrated in the eight main metropolitan areas of Milan, Naples, Rome, Turin, Genoa, Florence, Palermo and Bologna. Only 20% of the towns were growing, while the rest of the country, including several small and medium cities, was declining.

This trend changed in the 1970s, as it did in other Western industrialized countries, into a counter-urbanization phase when population growth rates become inversely proportional to the size of towns. In some cases it could even be called disurbanization, where the population of the larger cities declined. In the 1970s the number of Italian towns with active population balances rose to 59%. This was partly because of a further enlargement of metropolitan outer rings, but mainly due to the revitalization of smaller centres, spread all over the country, which were marginal in the former period.

Table 1.19 Housing stock in Italy by commune: increase in absolute figures and percentage ratio, by location (1961–91).

	Housing stock				Increase (%)			Increase (absolute figures)		
Location	1961	1971	1981	1991	1961–71	1971–81	1981–91	1961–71	1971–81	1981–91
Metropolitan areas[1]	1956385	2545998	2740879	2870084	30.14	7.65	4.71	589613	194881	129205
Medium-size cities[2]	974540	1244433	1452942	1534885	27.69	16.76	5.64	269893	208509	81943
Small cities[3]	496737	670149	813833	897260	34.91	21.44	10.25	173412	143684	83427
Rest of Italy	10786005	12973389	16845063	19500771	20.28	29.84	15.77	2187384	3871674	2655708
Total	14213667	17433969	21852717	24803000	22.66	25.35	13.50	3220302	4418748	2950283

1. Turin, Milan, Genoa, Rome, Naples
2. Brescia, Verona, Padova, Venice, Bologna, Florence, Bari, Taranto, Palermo, Messina, Catania, Cagliari
3. Varese, Como, Bergamo, Trento, Vicenza, Treviso, Udine, Parma, Modena, Ferrara, Perugia, Pescara, Caserta, Salerno, Foggia, Lecce, Cosenza, Siracusa, Sassari

Source: ISTAT.

Table 1.20 Non-occupied housing units, by location (percentage increase 1971–81 and percentage weight on total stock).

		Unoccupied in 1981
	Increase 1971–81 (%)	Total housing stock, 1981
Italy	103.8	19.9
Metropolitan areas	54.9	11.3
Urban areas	114.1	16.7
Scattered locations on plains	119.6	15.8
Scattered locations along sea coasts	156.7	35.5
Peripheral areas	111.5	29.5
Tourist areas:		
– total	119.6	48.3
– mature	115.8	52.2
– steady	85.7	39.3
– growing	152.0	45.9

Source: CRESME.

Table 1.21 Housing built per number of rooms and per average area (m^2).

	Average no. of rooms	Average area (m^2)
1984	3.99	87.3
1985	4.10	88.9
1986	4.13	90.8
1987	4.19	92.4
1988	4.19	92.6
1989	4.19	92.6
1990	4.18	93.2
1991	4.15	93.5
1992	4.13	93.6
1993	4.10	92.3

Source: ISTAT.

This deconcentration was related to the recession of central systems and the growth of peripheral production systems typical of the 1970s. More specifically, metropolitan systems, where large companies were located, lost jobs, especially in the manufacturing sector, and both in this sector and in the middle and low levels of the service sector, new jobs were generated in the smaller urban centres. At the end of the 1970s, a steep rise in high-skill jobs was observed in the great cities in the high-tech industries, research and development, corporate services, finance and management sectors.

The decline in population was thus attributable to innovative transformations that were taking place and that would eventually strengthen the position of the great city inside the urban network. At present, these new jobs are not enough to make up for the losses of lower-skilled ones in the more mature sectors. This might well change. There was already, in the late 1980s, a new flow of unskilled labour, including immigrants, as a result of the growth of services prompted by the new professionals.

The variation in time of the population deconcentration index for each region could thus be used as a clue to recent trends in the Italian urban network. The following typology was proposed by Guiseppe Dematteis (1992) and is derived from calculations made with the Lorenz index for the 1974–81 period:

(a) regions with a decreasing concentration index for the whole period
(b) regions where the inversion of the trend from concentration to deconcentration has taken place in the mid-1970s
(c) regions where this inversion appeared only at the beginning of the 1980s
(d) where concentration increased until the end of the 1970s and then remained steady
(e) regions where concentration increased for the whole period.

Table 1.24 shows the quite definite spatial distribution of the five types, from the North to the Centre and Northeast to the South. This distribution closely corresponds to the transformations of the economic structure.

Another useful indicator of the current processes is the number of corporate

Table 1.22 House buildings: number and volume (000 m^3) of building permits, number of housing units and number of rooms.

	Residential buildings							Housing				
	No. new	Vol. new	Vol. expansion	Volume total	New (%)	Expansion (%)	Total	Total housing	No. residential	No. rooms	No. annexes	Total
1981	50779	100310	10775	111085	90.3	9.7	100.0	247014	239328	1025791	832416	1858207
1982	53874	106616	11605	118221	90.2	9.8	100.0	265471	257732	1088315	853279	1941594
1983	51030	93200	10823	104023	89.6	10.4	100.0	230557	222656	940449	734865	1675314
1984	49308	88849	10671	99520	89.3	10.7	100.0	222417	214436	889283	709405	1598688
1985	44005	81775	9833	91608	89.3	10.7	100.0	200758	193494	814506	659481	1473987
1986	39752	71903	9084	80987	88.8	11.2	100.0	173361	166847	713803	581123	1294926
1987	39645	75466	8796	84262	89.6	10.4	100.0	191375	184530	802217	656633	1458850
1988	47340	90607	10948	101555	89.2	10.8	100.0	207834	200828	866211	713781	1579992
1989	46852	88248	10118	98366	89.7	10.3	100.0	196132	189594	821310	654861	1476171
1990	46211	91868	10801	102669	89.5	10.5	100.0	201857	194929	842716	668310	1511026
1991	46733	93213	11116	104329	89.3	10.7	100.0	204801	197978	850004	656801	1506805
1992	48734	95782	12110	107892	88.8	11.2	100.0	211526	204843	873581	674744	1548325
1993	43458	84151	11193	95344	88.3	11.7	100.0	188595	182688	772308	617175	1389483

Source: ISTAT.

Table 1.23 Constructions built per number of lodgings.

	Construct.	No. of lodgings	1 lodging (%)	2 lodgings (%)	3–15 (%)	over 15	Total
1980	56550	287044	50.79	18.83	24.09	6.29	100.00
1981	50779	247014	50.71	19.50	24.28	5.51	100.00
1982	53874	265471	50.52	19.58	23.96	5.94	100.00
1983	51030	230557	53.86	19.15	21.77	5.22	100.00
1984	49308	222417	53.87	19.43	21.71	4.99	100.00
1985	44005	200758	54.13	19.47	21.25	5.15	100.00
1986	39752	173361	54.73	19.48	21.10	4.69	100.00
1987	43525	191375	52.47	19.99	22.84	4.70	100.00
1988	47340	207834	51.27	20.19	23.90	4.64	100.00
1989	46852	196132	50.42	20.80	24.63	4.15	100.00
1990	46211	201857	49.17	21.04	25.20	4.59	100.00
1991	46733	204801	49.19	20.18	26.28	4.35	100.00
1992	48734	211526	49.47	20.38	25.90	4.25	100.00
1993	43458	188595	49.77	20.02	26.04	4.17	100.00

Source: ISTAT.

Table 1.24 Classification of Italian regions according to five clusters of urban trends.

	North	Northeast & centre	South
A	Liguria Lombardia		
B	Piemonte	Veneto Friuli V. G. Lazio	
C		Emilia–R. Toscana	Campania Puglia
D	V. D'Aosta	Trentino A. A. Marche	
E		Umbria	Abruzzi, Molise, Basilicata, Calabria, Sicilia, Sardegna

Source: G. Dematteis, 1992

services a city can offer. Of 36 core towns with more than 150 types of services, 23 are in the North, 7 in the Centre, and six in the South. But the main structural innovation is that, at the intermediate level, centres are much more evenly distributed, especially in the Centre–North. The "urban effect", once confined to a few great cities, has since the mid-1970s thus spread over most of the country. This is confirmed by the spatial patterns of the recent economic development, where a large range of activities have freed themselves from the need to be close to great metropolitan centres, though access to the great road networks is still very important. Specific local environmental factors are also becoming crucial.

Trends of agglomerations

The current spatial structure of the urban phenomenon may be outlined as follows: at the upper levels of the hierarchy there are a few great metropolitan areas that are either monocentric (Turin, Milan, Genoa, Rome and Naples) or polycentric (Veneto, Emilia-Romagna, northern Tuscany). The three dense southern metropolitan areas of Bari, Catania and Palermo are still large agglomerations, quite unconnected with their regional urban networks. Dense networks usually constitute the outer rings of the great metropolitan areas, whereas incomplete networks are usual in the most developed peripheral regions. There are finally some isolated urban centres, emerging mostly from the networks referred to above: Trieste, Ancona, Perugia, Pescara, Taranto, Cosenza, Reggio Calabria, Messina and Cagliari.The position of Italian cities in the European context is shown in Tables 1.25 & 1.26.

Inner urban development

The dynamics of metropolitan areas According to Dematteis'(1992) analysis of the urban life-cycle pattern in Italy, the most developed situation is found in the Northwest, which is mainly in the disurbanization phase, whereas the Northeast and the Centre are mainly in the suburbanization phase, with rings of peripheral towns still experiencing growth. Suburbanization is also found in almost all of the South, where it is more recent and where the growth rate of metropolitan areas is still very high.

Table 1.25 Range of the great European cities (DATAR–RECLUS).*

I	London
	Paris
II	Milan
III	Madrid
	Munich, Frankfurt
	Rome, Brussels, Barcelona
IV	Manchester
	Berlin, Hamburg
	Stuttgart, Copenhagen, Athens
	Rotterdam, Zurich
	Turin
	Lyons
	Geneva
V	Birmingham, Cologne, Lisbon
	Glasgow
	Vienna, Edinburgh
	Marseilles
	Naples
	Seville, Strasbourg
	Basel, Venice, Utrecht
	Dusseldorf, Florence, Bologna, The Hague, Antwerp, Toulouse, Valencia, Genoa

(*) The rankings are calculated by adding the positions each city occupies according to 16 different indicators: (1) population 1987; (2) population change 1970–85; (3) seat of multinational corporations; (4) infrastructure and technical activities; (5) engineers, managers, technicians; (6) research centres; (7) universities; (8) finance; (9) airport traffic; (10) port traffic; (11) culture; (12) fairs and shows; (13) conferences; (14) newspapers, magazines and books printed; (15) telecommunications infrastructure; (16) international level specializations (1989).

Table 1.26 Functional rankings of the main European metropolitan areas.

1. Global executive cities	
Pure	London, Brussels, Rome
	Amsterdam, Copenhagen
Complete	Paris Frankfurt Milan
2. Cities in positive technological/industrial transition	
Pure	Stuttgart, Turin
Complete	Munich
	Nurnberg
	Dusseldorf
	Cologne
	Strasbourg
	Hannover
Incomplete	Essen, Bordeaux
	Bologna, Tolouse
	Lyons, Duisburg
	Grenoble, St Etienne
	Bochum
	Dortmund
3. Cities in negative industrial transition	
Service cities	Dublin, Utrecht
	Liège, The Hague
	Rotterdam
Traditional ports	Marseilles
	Genoa
	Antwerp
4. Urban areas in structural crisis	
Functionally obsolete	Naples
	Edinburgh
	Glasgow
	Manchester
	Lille
Currently industrial	Birmingham
	Bristol
	Nantes
	Nancy

Source: Fondazione Agnelli 1989. Result of a factor analysis of 29 demographic, functional and structural indicators.

On the whole, data analysis for the 1971–91 period in the 36 largest urban areas shows the lower growth rate of metropolitan areas, both in comparison to the rest of the country and to the medium size cities (see Table 1.27). These facts are confirmed by more recent and refined data on the five great metropolitan areas (GMAs), Turin, Milan, Genoa, Rome and Naples, analyzed according to their administrative boundaries. In the 1971–81 period there was a decline in Genoa (-6%), a slight growth in Turin and Milan (2%), and a more obvious growth in Rome (5%) and Naples (9%). In the 1981–5 period there were negative trends in all three northern GMAs and growth rates similar to or higher than the national mean (1%) in Rome and Naples (1% and 4%). In all GMAs except Rome, the

Table 1.27 Population dynamics by province in the main levels of the Italian urban system, 1961–91.

	Residents			
	1961	1971	1981	1991
Metropolitan areas[1]	11208783	13478980	14075512	13887417
Medium size cities[2]	10011344	10741947	11214485	11355236
Smaller cities[3]	10559169	11151264	11909091	12215271
Rest of Italy	18844073	18764356	19367823	19320107
Total Italy	50623569	54136547	56566911	56778031

Notes: (1) Turin, Milan, Genoa, Rome and Naples. (2) Brescia, Verona, Padua, Venice, Bologna, Florence, Bari, Taranto, Palermo, Messina, Catania, Cagliari. (3) Varese, Como, Bergamo, Trento, Vicenza, Treviso, Udine, Parma, Modena, Ferrara, Perugia, Pescara, Caserta, Salerno, Foggia, Lecce, Cosenza, Siracusa, Sassari. *Source:* ISTAT.

Table 1.28 Employment dynamics by commune in the main levels of the Italian urban system, 1981–91.

	Employees		% increase
	1981	1991	1981–91
Metropolitan areas[1]	2560652	2428424	–5.16
Medium size cities[2]	1321264	1248636	–5.50
Smaller cities[3]	806707	781787	–3.09
Rest of Italy	15736975	15216125	–3.31
Total Italy	20425598	19674972	–3.67

Notes: (1) Turin, Milan, Genoa, Rome and Naples. (2) Brescia, Verona, Padua, Venice, Bologna, Florence, Bari, Taranto, Palermo, Messina, Catania, Cagliari. (3) Varese, Como, Bergamo, Trento, Vicenza, Treviso, Udine, Parma, Modena, Ferrara, Perugia, Pescara, Caserta, Salerno, Foggia, Lecce, Cosenza, Siracusa, Sassari. *Source:* ISTAT

core town has experienced a decline, both in the 1971–81 period and in the 1981–91 period.

In the past fifteen years the communes of both Milan and Turin lost some 100000–150000 residents to the surrounding towns and regions. Taken as a whole, the five metropolitan areas show a strong increase (20.25%) in population in the 1961–71 period, when the total number of residents increased from 11.2 million to 13.5 million. In the 1971–81 period the growth was a modest 4.43% and in 1981–91 there was a decline (–1.34%), which brought the total population back to 13.9 million people (Table 1.27). The absolute number of residents in the smallest cities ("Rest of Italy" in Table 1.27) increased by only 3.22% in the period 1971–81, whereas in the decade 1981–91 there was a decrease of 0.25%. As indicated earlier, the fastest growth of construction of housing units in the whole country occurred in the smallest communes. As the number of residents did not increase at the same pace, and even declined in 1981–91, this is solid evidence that most of the development concerned units used for tourist and leisure purposes and not as permanent homes. In the course of the 1990s there are signs that, on average, the decline in market prices in the real estate sector is affecting tourism locations more severely than the rest of the communes. This alone could be a strong deterrent against further construction of tourist homes, which are utilized on average only 50–60 days per year. However, the curbing of this phenomenon cannot be achieved only through market mechanisms, and it requires a national housing policy and local urban plans inspired by environmental values and sustainable growth goals.

The analysis of migration rates shows strong homogeneity. Negative trends were found throughout and in similar ranges (from the -3.5 per thousand in Naples to the -0.7 per thousand in Genoa), and significant variations among the different areas were found in natural population change rates. The former were usually inversely correlated with the latter. As for the internal dynamics, the strongly negative trends in all core towns are to be noted as opposed to migration balances, which are always positive in the outer rings.

Finally, employment trends in the 1971–81 period for comparable sectors of industry and the service sector, apart from industrial services and public administration, showed positive values only in Naples and Rome (19% and 15%). In the GMAs, employment in industry between 1981 and 1991 had a 3% fall, but with a net decline in the North and an obvious growth in the Centre and the South. In the service sector there was an 18% rise instead (with higher levels in Milan and Turin). Here again, balances were always more positive in the outer rings (province level) than in the core towns (commune level), but still not as in the rest of the extra-metropolitan region (see Tables 1.28 and 1.29).

Table 1.29 Employment dynamics by province in the main levels of the Italian urban system, 1981–91.

	Employees		% increase
	1981	1991	1981–91
Metropolitan areas[1]	4839423	4840838	0.03
Medium size cities[2]	3675187	3827171	4.14
Smaller cities[3]	4101172	4019009	–2.01
Rest of Italy	7809816	6987954	–10.52
Total Italy	20425598	19674972	–3.67

Notes: (1) Turin, Milan, Genoa, Rome and Naples. (2) Brescia, Verona, Padua, Venice, Bologna, Florence, Bari, Taranto, Palermo, Messina, Catania, Cagliari. (3) Varese, Como, Bergamo, Trento, Vicenza, Treviso, Udine, Parma, Modena, Ferrara, Perugia, Pescara, Caserta, Salerno, Foggia, Lecce, Cosenza, Siracusa, Sassari. *Source:* ISTAT

Although this analysis does not detect absolute concentrations of employees possible inside each GMA and in its core town, the trends very clearly indicate a relative specialization of the GMA against its regional context. In the 1981–91 period at the national level, there was an increase in service sector employees of little more than 1%, but in the five GMAs, the average global increase has been of almost 5%. The different parts of the country show different levels, with stronger specialization in the North (6.9% of relative incidence gained by the service sector in Milan, 4.7% in Turin), somewhat less in Rome (2.5%) and even a reverse tendency in Naples, where the industrial sector has seen growth of its relative weight in total employment (1.9%).

Inside the service sector, apart from industrial services and the public administration, which are not comparable, the main area of metropolitan specialization is credit, followed by transport and communications and wholesaling. The trend towards specialization is generally much greater in the core city, especially because of the credit sector, than in the outer rings. The only exception is the wholesale sector, which is much livelier in the metropolitan rings.

Peripheral urban development

Reshaping of urban networks in the last three decades has brought about major changes in employment, demographic, political–social and planning patterns. For the metropolitan areas they can be summarized as indicated below.

Relocation of the economic activities Large industrial plants have been abandoned and, with obsolete services and infrastructures, create the so-called "urban voids". They are seldom taken over by new industries, which prefer to go in the outer rings. Most of these areas, once marginal, are now in semi-central locations. They occupy high-value land but are also strategically placed for the redevelopment of the city, new infrastructures, neighbourhood regeneration, and so on. All this makes for tense conflicts, which the local authorities are often not able to mediate nor manage, with a paralyzing effect on the whole transformation process. Booming office jobs in the 1980s did not find an adequate supply of space and exert abnormal pressure on the central areas, evicting the residents, even in unsuitable areas.

Reorganization of major infrastructure New metropolitan functions require radical quantitative and qualitative changes in transportation and communications networks. Railways and urban highways, parking, port, intermodal interchanges, airports and telecommunications networks are always high on the public agenda, but what is implemented is still well below the requirements. In this respect, some Italian cities (e.g. Rome and Naples) are far below European metropolitan standards. Other cities have recently completed major infrastructure works, which enable them to participate in the European competition between cities and urban systems. Turin is probably the best example of the new positive attitude of local administrators. For example, during the course of the 1990s, this city has completed a brand new stadium, a new airport (including the only Alitalia cargo centre in the North of Italy) and the highway to France through the Frejus tunnel under the Alps. The work in progress, which is expected to be completed in 1999, includes: the intermodal link of *SITO* on 2.8 million m^2 (comprising an internal railway station for freight carriages); the freight platform *CAAT* dedicated to perishable goods, the new railway link for Turin, which entails the construction of 15 km long tunnels, with two and three levels in some points; and six underground stations, one of which will become the first Italian station for the high-speed train system anticipated for the next century. As of 1996, about 60% of the new railway link for Turin has already been completed.

Urban malaise The social spectrum is polarizing at the extremes. Next to the newly rich, pockets of poverty, marginality and deviance are growing, and often are the focus of aggressively organized crime, especially in the South. It must be noted, however, that the situation in the South is slowly but gradually improving since police action against organized crime was intensified in 1992 after the killing of Falcone and Borsellino. These were two of the most able judges, who were part of the "Pool Antimafia", a special unit to combat the Mafia in Sicily and elsewhere.

The housing market and troubled peripheral areas Polarization is mirrored in the housing market. Prices and demand grow faster (or decline more slowly) at the extremes of quality central or suburban houses and of substandard accommodation in derelict estates. Both facts, together with the demand for office space, reinforce the pressure on the central and semi-central areas. The market of low-quality housing – the bulk of the 1950s and 1960s expansion – is instead quite depressed. The risk is that progressive decay of these areas will follow, and the middle and upper-middle classes leave for suburban locations. There was a dramatic increase in land consumption per resident from 594 m^2 in the 1951–61 period to 2558 between 1971 and 1983.

Urban regeneration and decay Pollution levels are often high, green areas are few, natural elements go without care, and cultural life is virtually absent in many neighbourhoods. This is in opposition to the ambitions of the new urban classes. Urban regeneration is more and more often cited in public and private programmes, but mostly as a cosmetic appendage. People go for individual solutions, and among the consequences are the creation of segregated environmental enclaves and a sprawl of weekend and holiday houses in the countryside.

CHAPTER 2

Environment policy framework

2.1 The policy environment

The difference between stated public policies for the built environment and the outcome of their implementation is obviously a key issue in any country. In the case of Italy, attention must be given to a simple fact which is usually overlooked: there are many old laws which are still in effect, despite the fact that they date back from the pre-republican period and even from the nineteenth century, like the laws for the expropriation of private land for public purposes. Some laws have never been fully implemented. Indeed, there is a twofold level: on the first layer there is what would appear to be the planning system, according to a formal reading of existing laws, and on the second layer there is what is implemented in practice.

For example, in an audio cassette produced by the Royal Town Planning Institute (1991) it is maintained that one of the most important planning tools in Italy is the *comparto*, that is the allotment plan which is described by Article 23 of the Italian Physical Planning Act (see Ch. 3). It foresees the compulsory purchase of private properties in case a minority of property owners do not agree to participate in the development of a planned area. Indeed, this plan is rather important on theoretical grounds, for it gives a predominant role to the planned development against the lack of action by landowners. However, in reality, it has been applied only in a handful of cases in the past 50 years, so its practical role in the development of Italian cities has been negligible.

Thus, this section provides a brief description of the main public policies affecting urban land and property markets. Much more attention will be given in later chapters to the actual outcome of the stated policies.

2.2 Physical development policy and general housing policy

Even before the ruins produced by the Second World War it was felt that one of the key problems that the country would have to face after the fall of fascism was the reconstruction of its urban infrastructures. Obviously, after the destruction of war, quantity was much more appreciated than quality. In the postwar period and through to the 1960s, some 20 million Italians moved in search of a job, from South to North, Northeast to Northwest, and from mountain villages to the plains. The main social problem was a real quantitative lack of housing. This became one of the most serious social issues for about two decades.

The city master plans drawn up before the 1960s invariably had very optimistic perspectives: the key word was "expansion", as the population was always thought to increase, and even to double in the span of a decade. The role of urban planning was generally taken as the required condition to expand the built areas towards the countryside. The urban plans drawn in the 1950s and 1960s were invariably directed towards a steep growth of the number of residents, with housing to be built on areas used formerly as agricultural land on the outskirts of the built centres. It can be estimated that, if all the master plans had been implemented according to the maximum envisaged threshold, the Italian population should have increased to 100–150 million people.

A rent freeze was immediately taken soon after the constitution of the Republic, in order to achieve social peace in the country and to facilitate reconstruction. It was thought of as a short-term provision, whereas in reality the rent freeze was reapproved 22 times until 1978, when it was replaced by the Rent Control Act. Millions of households have been protected by the measure, and the most needy families (those who could not even pay the controlled rents) were entitled to a grant from a social fund set up within the Rent Control Act. However, the disappearance of a true rental market was the flip side of the coin, as the rigid yield cap imposed by the law discouraged private savings being invested in housing to let. Today rent control is limited to only a small section of the total rented housing stock, as the majority of the units are allocated on a pure market basis.

In the 1950s and 1960s, the prevailing answer to the widespread need for housing was a quest for more and more new buildings and, in parallel, a strong incentive for home ownership.

In practice the fiscal incentives were much less effective in determining the rate of home ownership than the tacit pact which was metaphorically signed between the growing middle class and the ruling party: no hassle with property taxes in exchange for political support. Starting in the early 1980s, however, the property sector has been the object of new taxes. It may not be entirely by chance that the first true sizeable tax on property ownership (the *ICI*, communal tax on real estate assets) was introduced in 1993, that is, in the middle of the transition started in 1992, which led to the collapse of the party system centred on the Christian Democrats.

In 1977 a new law introduced an important principle which has played a major role in the actual implementations of urban plans: any development activity can take place only if the concerned commune grants a "planning concession", which depends on the payment to the commune of a money contribution proportional to the marginal costs of urban services which are made necessary by the new construction and to the type of proposed building. This is explained in more detail in Chapter 3 and further referred to in some case studies (Chapter 6).

Between the end of the 1960s and the mid-1970s, it became evident that the housing needs of the population did not decrease, even when the quantitative output of the housing industry was higher than ever. Most of the new housing did not match the price range, the location and ultimate use which was at the base of the housing requirements of large sections of the population. In fact the statistics revealed that millions of new housing units had been built in tourist areas as second homes. Obviously, any increase in the output of this section of the housing market would not have any influence on the widespread housing needs in the large metropolitan areas. Two decades after the end of the Second World War, this was a sign, coupled with the burgeoning number of private cars, that Italy had reached a highly visible position (the "Italian miracle") in the restricted circle of developed western countries, with all the benefits and contradictions that go with it.

In the housing sector, the idea that the built stock should be the object of specific policies became a fact in 1978 with the approval of a new law. It provided funds for the restoration of the housing stock and for the upgrade of urban zones built with inadequate planning criteria. That law can be taken as the major turning point for property markets and for the urban planning profession. In the previous decades the key words were "new buildings on expanded urban land"; since the end of the 1970s the prevailing idea became "property restoration and inward development", mainly on vacant industrial areas. According to the new policy, both private owners and local administrations could propose and implement *Piani di recupero* (restoration plans), within the existing city master plans, taking advantage of state funds to cover a percentage of the actual costs. The range of works foreseen by the restoration plans went from ordinary maintenance of single buildings to urban planning rehabilitation of city zones. To implement the plans, the communes had the power to expropriate properties or to execute the restoration works within a plan and then claim the an appropriate proportion of the costs from the owners who had not joined voluntarily in the implementation.

In the case of restoration plans implemented entirely by private owners (usually grouped in cooperatives or consortia) or private developers, the communes could subordinate the approval of the plan and the issue of the planning permits to the allocation of a percentage of the restored units to certain types of households. In practice, in some communes out of 100 restored housing units, some 10–15 units had to be rented (at fair rent levels) to families below certain income levels, who had been previously evicted from private housing. This caused some problems in the housing market in which the restored blocks were located. The restored units were usually targeted for sale to medium- and upper-income level households, because of their central location and the good historical character of the renovated buildings. But many potential wealthy buyers were not willing to purchase at market price housing units where, in the same building or block, other households lived in the same type of units but belonged to the lowest income groups of society. It was a perfect example of how the best social intentions could not be achieved when the basic market rules and the tastes and preferences of end-users are ignored.

Under the 1978 law regarding restoration plans the government also launched the *Piano decennale* (the ten year development plan) which stated the goal of sustaining the private construction of dwellings. The actual numbers of houses constructed were low because only a few communes managed to prepare the required restoration plans in time to receive the available finance.

2.3 Social housing policy

The 1950s saw the vast INA *Casa* programmes; the construction of low-cost housing neighbourhoods, mostly in peripheral areas of large and medium-size cities. In the 1960s, with the novelty of the first government with a coalition including the Socialist Party, there were various attempts to mitigate the market approach to the housing problem. However, the quantitative idea was still there. A great impulse to build housing to rent to low-income households came from the 1962 law, which introduced new measures to expropriate private land for public use. This new social housing programme required all the leading cities, or those with a population greater than 50000 people, to allocate a portion of the city land to a plan for housing for low-income people (PEEP). The legislation enabled the communes to expropriate the land for these plans, which were soon renamed "167 plans". The communes were not always given the financial resources to actually implement the "167 plans", and certain areas remained "frozen" for years while no development (public and much less private) took place. Where the plans were implemented, the outcome was large-scale interventions, usually high-rise buildings located again in areas with low accessibility and poor urban services.

As the expropriated land was meant for public use, no profit-orientated activities were allowed within its boundaries. This proved to be an unforeseen social disaster because the developed areas were just rows of housing blocks, without any kind of retail and office activities. The nickname of *quartieri dormitorio* ("sleep-only neighbourhoods") was rather appropriate. However, it must be recognized that the "167 plans" were the last massive effort to provide social housing in the country. Since then, the stock of housing units for low-income people has remained more or less around 1.1 million units. As in other European countries, budget cuts in the institutions in charge of running social housing (the *Istituti Autonomi Case Popolari*, the autonomous institutions for social housing, replaced recently by the *Agenzie Territoriali per la Casa*, the regional agencies for housing) have stopped any expansion programmes. In the 1990s, these institutions have been allowed and encouraged to sell the public housing stock to the tenants and to the market at large. The government has issued certain economic and social indices which are meant to give priority to the long-term tenants in the public housing sector.

The financial resources to sustain public housing are collected through GESCAL (*Gestione Case Lavoratori*, the special fund for public housing for workers), set up in 1963 and closed in 1973. Collection is very effective: a tiny (about 1%) percentage is deducted at source every month from the pay of all Italian employees. This automatic deduction is barely noticed, being only a few thousand lire per month and one of several other deductions. In the end, however, the resources collected are huge, because they come from all those with a regular job in Italy, and also because the deductions from salaries are made on a regular basis, month after month, regardless of the actual spending on public housing programmes. Ultimately, the housing programmes for low-income households have been financed by the salaries of the workers at large. This method has been criticized, because it does not seek the contribution of the self-employed who, on the contrary, can qualify for a flat in a public housing scheme.

It must be noted that the deduction remained in effect even after the closing of the special fund agency (*GESCAL*) in 1973. The tax revenues so collected could be utilized for social housing, but there have been recurrent attempts by central government to divert them towards different uses.

2.4 Policy towards properties built without compliance to some urban planning regulations

The controversial *Condono edilizio*, that is the penalty remission for real estate properties not in line with planning laws, was introduced in 1985 amid sharp criticism from most professional associations of planners, environmental associations and left-wing political parties. It allowed the legal recognition of illegally built properties through payment of a fine. About 7 million applications brought to the state budget some 10 thousand billion lire. As described in this book, the Italian planning law system is highly complex and there are many laws that pretend to regulate the smallest details of property development. A window slightly larger than the approved designs in case of a housing construction or renovation, or an internal balcony transformed into a veranda without a licence, were all sanctioned as "illegal constructions". Even the opening (or the closing) of a door in an internal wall of a property was enough to label the entire property unit as *abuso edilizio* ("illegal building"), with the consequence that the sale of an "illegal building" could be deemed impossible until the "illegality" was removed. Or, if the sale had already taken place, the buyer could claim damages from the seller. Hence, in the overwhelming majority of cases, the penalty remission related to properties whose construction had taken place with minor flaws with regard to some planning laws or to some requirements established by a city master plan. In the most serious cases of illegal buildings, the law could not be applied. For example, constructions that had damaged the environment or had been built on public land, or non-authorized transformations of listed properties were all deemed not eligible for penalty remission. In all the rejected cases, as prescribed by the planning legislation, demolition was the only measure foreseen, albeit difficult to implement.

The great damage done by the *Condono edilizio* arose mostly out of the approval process of the law, rather than by what was written in its text. In fact, the debate on whether or not to have such a law started in the spring of 1983; the law was approved at the beginning of 1985. During the interim two years there were ample opportunities to actually commit the wrongdoings which would have been "forgiven through the payment of a fine and thus made legal by the new law. Hence the construction activity between 1983 and 1985 was sustained, to a certain degree, by a rush to initiate and complete as many real estate projects as possible, none of which would have received a regular planning permission.

The law brought large tax revenues to the state budget, but also gave the impression to the general public that any illegal building could be legalized through the payment of a fine. If there had been a first *Condono edilizio* there would sooner or later be a second one. In fact, in 1994 a second penalty remission in the property sector was introduced by the centre-right government led by the media and real estate tycoon Berlusconi. It was a populist measure to gain the electoral support of low-income people and to obtain an extraordinary tax collection, albeit lower than expected (about 4200 billion lire compared to an initial estimate of 10000 billion). That measure completed the environmental damages of the first penalty remission in the sense that it was only partially directed to writing off minor administrative offences.

The proponents of the first law had claimed to act in favour of low-income households who had committed negligible administrative mistakes (such as the opening or closing of a door in a wall separating two rooms of the same flat without the required permit) or who had built a house, without a licence, for themselves, and not for resale, especially in the postwar period. In 1994 this claim had much less foundation, if it was acceptable at all. Indeed, the law appeared to be tailored to the real speculative projects of housing and offices located on the outskirts of large cities. For example, the law admitted only properties of a maximum of $750 m^3$ of volume. But this was the single threshold of each application. Large condominiums built without planning permission could then be made legal through individual applications, each one below the stated limit. And entire property compounds, built in contradiction to planning standards, could be treated likewise.

It appears unlikely that the new centre-left government, elected in 1996, will even consider the enactment of a third penalty remission in the real estate sector. This is reassuring on the grounds of preservation of green areas and of the historic parts of Italian cities. But it seems necessary to go beyond the traditional two dimensions of planning: strict planning on paper and a much less than perfect implementation of plans. The "do-not" attitude which has inspired the urban planning legislation in the country needs to be replaced by a proactive vision, with new positive relations between public and private actors. Limits and restrictions can be successfully implemented when they make sense, when the controlling authority is truly determined to sanction at the outset any deviation and when the penalty is reasonable and proportionate to the crime, so that there is no need for the authorities to enforce the law on a discretional basis.

2.5 Policy for the retail sector

The retail sector in Italy is highly fragmented. It is estimated that there are about 750000 retail units, most of which are of tiny dimensions, in terms of both retail surface and enterprise budget. In the retail sector, central government policy has been designed and implemented on two layers. The first one, scarcely publicized and yet perfectly visible, is the general fiscal policy, which has been for decades more than generous towards the retailers. In exchange for the traditional voting support to the coalitions in power (led for decades by the Christian Democrats), the retailers have been allowed to be one of the major sources of tax evasion in the country. In the 1970s and 1980s, minor retail activities also played an important role in absorbing thousands of workers dismissed by the large industrial firms. On the second layer, there is the policy directed openly to the retail sector, with the stated goal of modernization of this branch of the economy.

Modernization was the long-term objective of the first law to regulate the development of the retail sector, which was enacted in 1971 (*Disciplina del commercio* or Retail Act). "Modernization" meant the increase of the average area of the retail outlets, by setting minimum surface standards, differentiated by branch of business, for granting new licences. But it has also meant the harmonization of the retail activities accomplished in *sede fissa*, that is, in a real estate unit, with the retail activities conducted in open-air markets, whose presence in the Italian cities has deep historical roots and is still rather important. To this end, the law required each commune with a population above 30000 people to prepare a *Piano di sviluppo e adeguamento della rete di vendita* (a plan for the harmonization and development of the retail network). What is perhaps the key point of the retail policy is the requirement for each commune to divide the entire territory in commercial areas and fix an overall upper limit for the shopping area of the commune, for each of the various branches of retail activities. This limit (the famous *contingente*, or "cap", on retail outlets) is then used as a guiding rule to grant new licences on a branch-by-branch basis, or to refuse new ones if the limit had been already reached.

The regions have full responsibility to formulate and monitor policies with regard to the development of shopping centres. Generally speaking, the attitude of the regions has been to draw comprehensive plans for the location of the truly large shopping centres, that is, those comprising an internal hypermarket, which required a specific authorization (the famous *Tabella VIII*). With regard to the shopping centres, there responsibilities are split between communes and regions. The communes are responsible for granting the planning permits required to build the structure of the shopping centres, whereas the regions have the power to grant the commercial licences for the hypermarkets with large surfaces (above 1500 m^2), which are the true engine of the large shopping centres. This division of responsibilities has been the cause of some inconsistency in the action of local administrations. In the 1990s several regions in the North and Centre have approved general policy rules that will make the issue of permits for large shopping centres increasingly strict. The number of existing centres already appears to be very close to saturation in the most wealthy regions (for example, Emilia-Romagna, Lombardy, Piedmont, Veneto), so the good planning intentions are welcome, but they come rather late.

Italy is considered a country with good growth potential for shopping centres. As of 1996 there are about 200 fully developed shopping centres, mostly in the North and Centre. Recently, the associations of retailers e.g. CONFESERCENTI,

ASCOM) who are rather powerful from an electoral viewpoint, have increased their pressure on the central government to put a national ban on opening new large shopping centres. It seems highly unlikely that the government could introduce measures to limit competition within the European market. Thus, if any measure is to be taken, it will have to cover international chains as well as national ones.

Nonetheless, it should be remembered that the approval process is entirely in the hands of the regions, so the situation is rather diversified. The general attitude of local administrations (communes and regions) towards large shopping centres has moved from an unconditional welcome to one of severe scrutiny. There is a growing awareness that the location of a shopping centre could bring heavy environmental costs and that an excessive number of shopping centres can greatly reduce the quality of life in the urban centres.

2.6 Policy for the environment

In Italy, environmental policies are the responsibility of the Ministry for the Environment, which was set up in 1986. The Ministry has jurisdiction over the following issues: air, water, waste, noise, environmental impact analysis, regional planning, industrial risk, parks and protected areas. The jurisdiction on many of these items is shared with many other ministries, notably the Ministry of Cultural Assets, which is responsible for landscape planning and cultural heritage, and the Ministry of Commercial Fleet, which is responsible for sea parks.

Relevant environmental planning tasks are assigned to the regions, which have been in existence since 1970 and are responsible for regional planning and for the implementation at the regional level of environmental policies based on EU directives and key national laws. The recent law (1990) has also transferred competencies on regional planning to the provinces, which have long been empowered to monitor compliance with environmental norms.

Since its establishment in 1986, the Ministry for the Environment has reduced the time required for the integration of EU directives into national law and their implementation, in which Italy was previously lagging behind. The Ministry is responsible for EU directives on environmental assessment (which have been applied in part only), on major industrial risks (the *Seveso* Directive, 1988), and it has updated legislation on air, waste, drinkable water and seawater pollution, accepting European directives on these issues.

In practical terms, the Ministry has worked mainly on emergency situations, intervening in areas of environmental risk, contaminated land, on waste illegally discharged or illegally exported abroad, and with extraordinary interventions for the Po Basin and the Adriatic Sea, which are subject to intense pollution.

As the legislative activity grew, budget allocations increased accordingly. Expenditures for the environment grew, in constant values, from 4484 billion lire in 1981 to 5300 in 1988 and to about 6000 in 1991, with an increase above 6% per year. Nevertheless, actual expenditure grew more slowly because of difficulties of public administration in implementing plans.

Environment and regional planning

The two most relevant laws on environmental policy that are of significance for regional planning are:

- *The Galasso Law (1985)* It includes urgent norms for the protection of areas of particular environmental sensitivity: sea-coasts, banks of rivers and lakes, mountains above a certain height, forests, and other specified individual areas. The regions have to prepare their "landscape plans" or integrate the landscape dimension in their own territorial plans. These must set out the rules for the area and the use of the protected areas and protect other landscape values.

Only a few regions (Emilia-Romagna, Liguria, Veneto, Abruzzo, Basilicata and Marche) had by 1995 approved landscape plans in the terms laid out by the law.

- *The Basin Plan (1989)* This has the aim of assuring the defence of the earth, and hydrogeology, the quality of the water and the correct management of the water property for different uses. It also has the tasks of defending the general environment of the water basins.

A qualifying aspect of the law is the setting up of the Basin Authority, which has the task of assuring the planning of the basin considered as an ecosystem, coordinating the different uses of the water resources.

Water

The principal law governing policies for the quality of the water in Italy is that of 1976 and its subsequent amendments (also known as the Merli law). It is partly superseded, but establishes the parameters of water depletion, both of surface water and that from underground, both public drainage, supplied or not with filters.

Standards for the quality of the water, and legislative indications of the quality of bodies of water, are, however, missing for the text of the law. The regions, although in theory having the power of defining the parameters of the quality of the water bodies, have usually decided to renounce it. The only exception is that of Lombardy, which has classified its rivers on the basis of the quality of the water and its intended uses (in the order: for drinking, for swimming, for agricultural use and for industrial use).

The main EU directives have been received with the aim of establishing a standard of water quality for some specified uses.

The EU directive on drinking water has been the object of readjustment because of the pollution via pesticides in the water wells, which have exceeded the maximum limits for some substances, for which provisions have been granted to many communes in the lowlands of the River Po.

Air

The law on air pollution is the oldest of the Italian environmental laws, dating from 1966. The law foresees the division of the country into three zones (A, B, and C) with different content levels of emissions from civil thermal plants. It sanctions industry's obligation to "adopt such devices to contain polluted emissions within the narrowest limits which technological progress allows".

In 1977, responsibility for levels of atmospheric pollution was partly handed over to the regions and the provinces. As regards the moving sources (which represent one of the principal problems of today) all EU directives have been received

For fixed sources of pollution, the Minister of the Environment has been implementing the EU directive of 1988 and the Helsinki Protocol. It gives objectives for reducing global emissions for the large existing combustion plants. For new plants, however, new limits are established for individual factories.

The most recent laws have strongly reformed legislation on atmospheric pollution, because they require that the regions adopt air-quality levels on the basis of guide values established by the Minister of Environment. Likewise they bring into consideration the authorization of all sources of atmospheric emissions.

Refuse

The current law controls the disposal of all types of refuse, detailing also the relevant principles for the reduction of the refuse as well as the quantity and the danger for the various collections. It introduces the principal that foresees the disposal of urban refuse for the communes (also in private concession), whereas the disposal on industrial refuse is placed in the hands of private contractors. Moreover, all those who operate in the refuse field (collection, transport, storage, treatment and disposal) have to be placed under the regional authorization. The major part of the capabilities with regard to refuse are placed in the hands of the regions, yet evaluation of the environmental impact for the disposal of toxic and harmful refuse is a matter reserved for the state.

Two laws at the end of the 1980s have clarified some key aspects to reduce and recycle refuse. The first sets norms for the different categories (harmful urban refuse, rubbish from fruit and vegetable markets, glass, paper, plastic and aluminium). The second promotes consortiums, on a regional and interregional scale, for the industrial recuperation of the secondary materials. The laws also dictate laying norms to remove the big locational difficulties that the refuse disposal plants are meeting in the whole of the country, because of the opposition of the local population.

Noise

The ruling applied to noise pollution in Italy is relatively recent. The Minister of the Environment issued a decree in 1991, which established maximum limits on the emission of noise in housing areas and in the external areas. It has not in fact been put into force, as there has been a moratorium for five years for the productive plants and for three for all the other subjects.

The maximum emission limits are shown in Table 2.1. Classification of areas has to be carried out by the communes. Regions foresee and dictate the directives for guiding the communes for the restoration of the zones in which the limits of

Table 2.1 Noise emission levels.

	Max. dB day	Max. dB night
Particularly protected areas	50	40
Mainly residential areas	55	45
Mixed areas	60	50
Intense human activity	65	55
Mainly industrial areas	70	70

the law are exceeded. Companies, who do not have duties under the law, have to present an adequate plan within six months, which has then to be put into action within 30 months.

2.7 General policies affecting the cities

Traffic and parking

In Italy the greatest share of freight services is undertaken by road transport. As private traffic has constantly increased, the construction of new highways has been banned for years under the pressure of environmental concerns and also to curb public spending. The result is a highway network with insufficient capacity along the most busy routes and at certain periods of the year. The railway service before the end of the 1980s had been losing business clients for decades. Since the railway company *Ferrovie dello Stato* was incorporated (i.e. took the legal status of a private firm, SPA) there has been a positive effort to increase its productivity and to convert vacant land and properties all over Italy.

In 1989 the government enacted the "Tognoli Law" which creates general rules concerning the construction of underground car parks in the main urban centres. Also at surface level there have been notable changes with the introduction in the major cities of *Zone a traffico limitato* (ZTL, restricted traffic areas), and *Aree pedonali urbane* (urban pedestrian zones), anticipated in the new *Codice della strada*, the Highway Code, issued in 1992. The city of Florence, for example, in 1996 extended the ZTL to an area of about 3 million m^2, within which there is a pedestrian zone of 250 000 m^2, the largest in Italy, plus other smaller zones making a grand total of about 500000 m^2 reserved for pedestrians.

Decentralization

For many years, decentralization of central government (and its functions) has been debated. This idea is the most visible aspect of a proposed policy of decentralization of powers, which is in accordance with the present Republican Constitution. It might result in the strengthening of the autonomy of the local administrations (first of all, regions and communes), and ultimately in the decision to introduce in Italy a federal form of government, similar to other Western countries (e.g. Germany). A practical example of the ideas which have been advanced is given by research published in 1995 by the *Fondazione Agnelli* (the social and economics studies foundation of *FIAT*). Its ultimate proposal is a *capitale reticolare*, that is a "network capital". In practice, some ministries should be relocated from Rome to the main cities, both in the North and the South. The same should happen to major public firms such as *RAI* (television), national institutions and international bodies with offices in Italy. It is maintained that this relocation would reduce congestion in Rome and, above all, would benefit economic growth and the diversification of the economic base in the chosen cities.

Currently the company *TELECOM Italia* is implementing a plan to lay down optical fibres in all the major Italian cities. In addition *Ferrovie dello Stato* is improving the intercity service and has advanced plans to build the first two lines of the high-speed rail network of Italy, (north–south from Milan to Rome and Naples, and west–east, from Turin to Venice). These two major infrastructures would greatly enhance the communication facilities of the country and might eventually be decisive factors in effecting the relocation of major capital city functions to the peripheral cities.

PART II
THE URBAN LAND MARKET

CHAPTER 3

The framework within which urban land and property markets function

3.1 The legal environment

Laws, acts, responsibilities and plans

The main Italian Physical Planning Act is the *Legge urbanistica generale* (1942) which regulates urban and regional planning in the whole country. It is easy, at present, to find flaws in this law, as it is basically inadequate to cope with the needs of an advanced society and a complex economy such as that of 1990s Italy. However, at the time of approval, this law was quite appropriate to face the urban problems of the times, which were primarily of population growth and territorial expansion of cities. The law was also innovative, as it had foreseen problems such as urban renovation and restoration, which were put on the agenda of local government policy-makers, in the 1970s and 1980s, well after the reconstruction phase. By the end of 1995, there was growing support among Italian planners for the idea that the law should be replaced by a new one, rather than amended here and there. However, it is impossible to predict when this radical change will take place.

This section will present the main features of the law, as well as the key points of the proposed reforms. It is worth emphasizing that some of the features of the law have never been implemented, and this will be indicated in the following description. Nevertheless, the non-implementation of some parts of the law should not be dismissed superficially. Rather, there are deep-rooted reasons, both in the text of the law and in the society, which prevented its full implementation.

The law called for a top-down approach, had a highly centralized vision of urban planning and local government, and gave no consideration to the time factor. Each plan at the lower levels of government could be designed and implemented only when the corresponding plan at the next level up was in place. The non-approval, or delay in approval (which had the same result), of any given plan would bring the planning chain to a halt. The rationale behind the law was that urban planning had to restrain the physical growth of cities, and that such growth would occur anyhow, so that the local authorities' role was to put order in the development process. In other words, the public authorities had to preserve the common wellbeing of society from the private search for profits and windfalls in the land and property markets. In the final part of the book, the main goals and planning tools of the physical planning act will be put under close scrutiny, as they seem inadequate for today's cities, where the key problem is no longer physical expansion.

Before taking a detailed look at the provisions of the Law of 1942, it is helpful to recall that the main planning authorities in Italy are the state, the regions (*regioni*) and the communes (*communi*). As mentioned above, the recent reform of local government also gave certain planning powers to the province, albeit limited and yet to be tested in practice. The state directs and coordinates the administrative activities of the regions having an ordinary statute. Furthermore, the state can set out planning guidelines for national sites, paying special attention to balanced regional development, national interest and any potentially serious environmental impact. The state can also determine national planning standards. Another function typical of the central level of Italian administration is that decisions on the location of public works to be implemented either by the state or on state land have to be taken in accordance with the region concerned.

The regions have extensive administrative authority and, above all, wide legislative powers in the field of urban and regional planning. In this area, together with the other domains indicated by Article 117 of the Italian Constitution, there can be two-tiered legislation:

- a national law, issued by Parliament, setting out general guidelines
- a regional law, issued by a particular region, setting out regulations that must not clash with either national law or any other regional law.

Each region has its own administrative structure which varies from region to region, although the regional administration is usually composed of departments. In the field of rural and urban planning, the responsibility is of the *Assessore all'Urbanistica*, regional councillor for urban and regional planning, who holds administrative and legal powers. With the new electoral law of 1993, this councillor is chosen by the president of the region, who is elected directly by the voters along with the members of the proposed regional government.

The main implementation tools are composed of a long list of different types of plans, the main ones being:

- *Piano regolatore generale comunale* (PRGC), the physical master plan of the city
- *Programma di fabbricazione* (PF), the development plan
- *Piano particolareggiato* (PP), the detailed plan
- *Piano di ricostruzione* (PR), the redevelopment plan.

Apart from the PR, which was a planning tool devised for the reconstruction of Italian cities damaged during the Second World War, the other planning tools continue to be applied. In addition, there are also other categories of plan which have remained on the books but which have been rarely if ever implemented in actual planning practice. The main example of this is the *Piano regolatore generale intercomunale* (PRGI), a PRG whose coverage was expected to encompass all the communes that compose a metropolitan area. The most renowned example is the *Piano intercomunale Milanese* (PIM) of the 1960s, which never reached an implementation phase and whose difficulties epitomize the difficulties of urban planning in Italy. There is also the *Piano territoriale regionale* (PTR), a regional territorial plan, which remained totally forgotten until the early 1970s, when the Italian regions were set up, as described in Chapter 1.

By contrast, the *Piano di lottizzazione* (PL), the plan for subdividing land into plots, has been used widely by the communes. It should be noted that quite often the PL takes up different names and acronyms in the Italian regions, as this type of plan must respond to the regional urban planning acts and also to norms contained in the master plan (PRG) at city level. However, the substance does not change and it consists of standards and procedures to be followed when a large land property is subdivided into smaller projects.

Until the 1960s, the law of 1942 was only partially implemented. It has been implemented widely, albeit not in all its elements, only since the mid-1970s, so that it is now possible to evaluate it fully.

According to Articles 5 and 6 of the Physical Planning Act of 1942, the commune is where urban planning is actually implemented. The communes have a fundamental role in the Italian system, and land use in any given location is generally defined by one of the many planning procedures that the commune concerned can utilize. These include the following. At the general level is the *Piano regolatore generale comunale* (PRGC), the physical master plan of the city (the acronym PRGC has been replaced, in ordinary language, by a simple "PRG", because there is no risk of confusion. For the communes with a few thousand inhabitants, the law requires, instead of a full PRG, only a *Programma di fabbricazione* (PF), a development plan, which is a sort of simplified PRG that can be implemented directly, without the need of lower tier plans such as the detailed plans (see below).

Other kinds of plans include:

- the *Piano particolareggiato* (PP), the detailed plan
- the *Piano di lottizzazione* a plan for subdividing land into plots
- the "low-cost housing plan" (*Piano di edilizia economica e popolare*, PEEP)
- the "light industrial plan" (*Piano di insediamenti produttivi*)

The commune can authorize "multi-year implementation programme" (*Programma pluriennale di attuazione*, PPA) usually covering a three-year period. The commune can also enact a "city building code" (*Regolamento edilizio*) and "health and hygiene code" (*Regolamento d'igiene*). Finally, the commune has the power to grant building licences and planning approval and obviously has the power and duty to monitor all development activities.

The various planning acts and their order of importance

According to the Physical Planning Act of 1942, planning activity can be implemented at the following three levels: region, commune, and subcommune. These levels correspond, respectively, to area plans, communal plans and executive plans. These plans have different influences and roles according to the size of the territory covered.

Thus, one can identify, in the text of the law if not in the planning practice, the following three kinds of plans:

- *Piani d'inquadramento*, or *Piani-quadro* These are general context plans or master plans which define the objectives and programme guidelines for a large area, usually a region or a large part of a region. According the 1942 law there should be a *Piano territoriale regionale* (PTR, a regional territorial plan), the *Piano territoriale paesaggistico* (PTP, a landscape plan), and the *Piano di comunità montana* (PCM, a mountain community plan). The general planning law also requires a *Piano territoriale infraregionale* (PTI), a plan for areas spanning over more than one region.

- *Piani generali*, or general plans These define, in accordance with the guidelines of the structural plans, the land-use of a commune. The Italian acronyms for some of these kinds of plans are as follows: *PRG*; *PRGI*, a plan whose coverage includes the communes of a whole metropolitan area, and *PF*, the development plan.
- *Piani attuativi*, implementation plans These usually concern small areas within a commune and they define, in accordance with general plans, the physical changes allowed in those areas and how they should be carried out. The Italian names and acronyms for some of these kinds of plans are as follows: *PP*, detailed plan; *Piano di zona* (PZ), a local plan for low-cost housing; *Piano per insediamenti produttivi* (PIP), a plan for artisans and light industrial buildings; the *PL*, a plan for subdividing land into plots, referred to above; the *Piano di recupero* (PR), the urban renewal plan, and so forth.

Plans and planning coverage

The *PRG*, the physical master plan, is the key planning facility available to a commune administration. Through the *PRG* a commune can set its objectives and guidelines concerning development, change and preservation. According to the 1942 law, as modified and updated by subsequent laws, the domains of the *PRG* include the following:

- communications networks (roads, railways, rivers) and associated infrastructures
- subdivision into zones of all land within the commune boundaries; indication of any expansion zones for building activities; defining development restrictions and regulations for each area
- areas reserved for public use, for state buildings, or for other developments of social need
- density ratios, whose minimum figures were defined at national level by decree in 1968 (known as the decree on "urban standards" or briefly on "standards"), and subsequently by specific regional laws
- development restrictions on areas of historical or environmental value, and on areas having special landscape characteristics
- planning regulations which must make explicit what is set out in the development plans, and administrative procedures for the implementation of the *PRG*.

Planning process for local plans

All Italian communes are entitled by law to draw a PRG, the physical master plan. However, the communes included in special lists compiled by the regions are also obliged to draw up a PRG. In several regions such lists comprise all the communes located within the regional boundaries. The communes, which are exempt from the obligation to draw a PRG (usually the small villages), must draw up a development plan (*Programma di fabbricazione*), which is a planning document similar to a PRG but is more restricted in its contents and simpler in its procedural requirements. In 1995, it was estimated that at least 90% of all Italian communes have either a PRG or a PF, and all the provincial head cities and the other major communes have a PRG.

The process that leads to a PRG is a complex series of administrative decisions, legal requirements and writing and drawing activities, which are summarized below:

1. The decision of the "city council" (*Consiglio comunale*) giving the go-ahead for a PRG.
2. Drawing up the PRG, its "adoption" and publication. The plan is "adopted" by the city council and, once the "adoption act" has become effective (that is, after a short period of time), the PRG is registered with the commune secretariat, where anyone can examine it during the following 30 calendar days.
3. "Safeguard measures" are issued by the city council as soon as the PRG is adopted, as prescribed by the 1952 Law. It must be noted that a decision in 1984 of the State Council (*Consiglio di Stato*, which is totally independent as indicated by the Constitution) sanctioned the absence of any time-limit on the restrictions imposed on the use of privately owned land which would entail either full expropriation or a total ban of development. In practice, a commune could no longer place on privately owned land an unlimited ban on development in anticipation of the construction of a public facility such as a school. The ban had to be linked to a time limit (five years), after which, in the absence of public action with regard to the construction of the public service, the restriction had to be cancelled.
4. Observations on the proposed PRG can be submitted by anyone within the 30 days. Thus, the PRG is subject to modification from the very date of its adoption by the city council.
5. Responses and specific recommendations are given by the city council on all observations presented within the time limit.
6. The proposed PRG, together with all the observations and alterations, is sent to the regional authority for approval. The region has to check the legality of the proposed plan and ensure that it complies with the 1942 law, which requires that:
   - it does not conflict with any existing regional territorial plan
   - there is a sensible siting of public buildings of national interest, together with their services and infrastructure
   - full landscape protection and preservation of buildings and sites of historical, monumental, environmental or archaeological interest is ensured

- urban standards are maintained.

7. The *PRG* is approved by the region through its' own decree, which is binding on both the public's observations and any subsequent recommendations. The *PRG* becomes effective from the moment it goes back to the city council and is registered with the department concerned.
8. Re-publication of the *PRG* can be necessary whenever changes in planning policy are proposed during the approval process. These changes can be proposed either at the city council level, when the counter-arguments are prepared in response to previous observations on the *PRG*, or at the regional council level, when the *PRG* is examined for approval. In many cases, this process leads to a second adoption process for the *PRG* and ultimately to a republication of the *PRG*. The whole process takes a minimum of two years, and even up to ten years or more to complete in case of large cities.

Once the *PRG* is approved, there are no time limits on its legal validity, as indicated by the 1942 law. However, the restrictions on building rights, which entail compulsory purchase in the public interest, cease to be valid five years after the date of *PRG* approval. This is the case when the restrictions are imposed on private properties in order to implement the planning objectives of the *PRG*, such as construction of roads, schools and so on. These restrictions, which impose a total ban on development, are still valid when used to enforce a no-go area along roads, railways, graveyards and so on. The same is true for restrictions that cover areas where accepted land uses are very limited because of landscape protection measures or hydrogeological, historical or architectural interests.

There is no limit to the number of amendments that can be introduced to an existing *PRG*. The process of formation and approval of amendments is the same as described for the first approval of the *PRG*.

Traffic and parking plans

The new *Codice della Strada* (Road Code), issued in 1992, introduced a new complementary planning tool called *Piano urbano del traffico* (*PUT*), the city plan for traffic control. From April 1993 this new plan had to be introduced by communes with a population more than 30000 people and also by smaller ones visited by many tourists. As of 1996, most communes have actually approved a *PUT* and are in the implementation phase. From a city planning viewpoint, the main feature of *PUT* is the power given to communes to name *Zone a traffico limitato* (*ZTL*, restricted traffic areas), and *Aree pedonali urbane* (urban pedestrian zones), in any part of the city, usually the central areas. In addition, the communes were given powers to extend paying parking areas (the "blue zones") within the restricted traffic areas and in the zones adjacent to monuments, and of special interest. The goal of this norm was to ameliorate traffic flows, reduce pollution and enhance urban quality of life, especially in the congested central areas of the city. In mid-1996 an amendment to the Road Code has given the communes the power to put in place automatic means of payment of "entrance fees" at the border of the *ZTL*s.

The *PUT* has to be designed and implemented in accordance with the *Piano urbano dei parcheggi* (*PUP*, the plan for urban parking spaces), which was introduced in 1989. According to the legislation, the 16 largest Italian cities (Rome, Milan, Turin, Genoa, Venice, Trieste, Bologna, Florence, Naples, Bari, Reggio Calabria, Messina, Cagliari, Catania, Palermo) had the responsibility of approving a *PUP* within six months of the enactment of the law. In addition, the regions could issue special lists of smaller communes that would also be required to draw up a *PUP*. The incentives to approve a *PUP* were given by the fact that only the communes with a *PUP* would be entitled to receive state finance for the construction of *Parcheggi di scambio*, intermodal parking buildings, and *Parcheggi pertinenziali* (underground parking buildings) whose purchase is reserved for property owners of nearby areas.

So far, the construction of new parking facilities has not had a sizeable impact on traffic conditions in the major cities. The most effective measures seem to be the *PUP* and the designations of pedestrian areas and *ZTL*s. These are tools that require more human resources than financial assets, so their implementation is well within the capability of most communes. At the beginning of the project, most of the owners of retail properties were against the restrictions (as they are against any innovation in traffic rules for fear of losing clients). After some time, the traffic plans of the new administrations elected in 1993 proved to be rather effective in reducing car pollution generated by crawling traffic, and there are several cities where retailers in central zones, not originally included in the blue zone, asked the respective commune to extend the *ZTL* to their area.

As a result of growing support for urban ecology, a national law was approved in 1991 to finance the construction of urban pathways reserved for bicycles. The law enables (but does not oblige) communes to approve a *Programma delle rete ciclopedonale* (*PCP*), a programme of the road network for bicycles and pedestrians. It has to be designed within the *PUT*, the city plan for traffic control, and made consistent with the *PUP*, the plan for urban parking spaces. Some regions have approved regional laws that aim at coordinating the plans for bicycles routes at provincial and communal levels. Once again, there was a rational approach (through various tiers of planning) to a new problem (the bicycles versus other means of transport) with inadequate attention at the implementation phase. In fact, very few communes have actually built bicycle routes so far, the main reason being a lack of funds.

Plans for the retail network

At communal level, another complementary planning tool is the *Piano di sviluppo e adeguamento della rete di vendita*, the development plan of the retail network. This planning tool was envisaged by the main law in the field of retail planning, which dates back to 1971 and whose purpose was to rationalize the sector and increase the efficiency of many small businesses, usually run on a family basis.

The communes have usually failed to establish rational links between the development plan for the retail network, the urban traffic plan and the city master plan. The opening of large shopping centres, often in small communes just outside main cities, has been a major source of traffic nuisance and pollution in the surrounding areas, and a serious threat to the viability of the retail network in the central part of the city.

It must be noted that in Italy the large-scale shopping centres (those comprising a hypermarket plus a gallery and some specialized stores) must obtain one licence to operate in the retail sector and a second licence to build the physical facilities in specific locations. The two licences are issued by two different local administrations: the region, in the case of the commercial licence, and the commune, in the case of the planning permit. On one hand, this division of the authorization process is perhaps one of the causes of a general failure to blend the new facilities into the ordinary urban fabric. On the other hand, the small communes usually chosen as a suitable location by the large retail outlets have proved to be completely unarmed, with their usual obsolete planning tools, to preserve green land from a rapid growth in high-discount outlets and large shopping centres.

Nevertheless, when compared to the USA and even to most European countries, Italian local government has, on paper, a wide range of planning tools to curb and steer large retail outlets. What has been lacking, perhaps, is a true cooperation between public administrators at all levels (commune, province and region) in order to design a practical policy for the retail sector, in the search for common ground between residents, owners of small retail outlets and promoters of shopping centres.

In the mid-1990s, some of the most active regions in the North (for example, Piedmont and Emilia–Romagna) have approved regional laws that are severely restrictive towards the opening of new shopping centres. However, it seems that these new measures have been formulated rather late, as they have been targeted towards future new applications, whereas the many projects already in the pipeline have some chance of approval. The number of existing shopping centres and discount outlets is already considered too high by a variety of interests, from the owners of small stores to the political parties and opinion movements closer to the ideas of sustainable urban growth, and to the owners of the shopping centres already in place. In addition, it seems that in several cities real estate prices in the zones immediately adjacent to the shopping centres have been negatively influenced by the increased traffic and noise.

Private law relating to land

These issues are regulated by the Italian Civil Code (*Codice Civile*). Land and property transactions since 1989 can be undertaken only by intermediaries listed in special registers held at the local chamber of commerce (see Ch. 4).

Implementation of plans

According to the 1942 Law, the planning process at the commune level has two phases: general planning through the *PRG*, the physical master plan, which sets overall objectives and has no time limit (except as described above); and executive planning through the *Piano particolareggiato* (*PP*), the detailed plan that has the role of putting into practice the objectives of the *PRG*.

According to the general planning law, the *PP* must have a validity limited to 10 years. Within this time limit, the plan has to be implemented and the foreseen expropriations must take place. The parts of the plans which are still not implemented at the expiration of its validity period must be the object of a new *PP*. When the validity of a *PP* is expired, only the street alignments and the zoning indications remain in effect.

On the one hand, it should be noted that in Italy only a few detailed plans have been prepared, approved and implemented. On the other hand, several other planning procedures have been introduced by the national legislature which have become de facto "detailed plans". These kinds of plans include the following:

- *Piano di edilizia economica e popolare* (*PEEP*), the plan for low-cost housing, which was introduced in 1962 and has been subsequently modified and integrated
- *Piano di insediamenti produttivi* (*PIP*), introduced in 1971 and relates to artisan, industrial, commercial and tourist activities
- *Piano di lottizzazione* (*PL*) the plan for subdividing land into plots, introduced in 1967 and regarded as an alternative to the detailed plan, – it can be proposed either by the local public authority or private owners
- *Piano di recupero* (*PR*), defined in 1978 and concerning urban restoration and renewal.

Any kind of plan having the function of a "detailed plan" cannot include regulations that conflict with the *PRG*. The terms of reference of the detailed plans are restricted to issues that are not specifically addressed by the physical master plan of the city. The *PEEP*, the plan for low-cost housing, has been the planning procedure most widely implemented by local authorities to address the housing question for the poor in Italy. When promoted by private owners, the *PL* (the plan for subdividing land into plots), sanctions an agreement between a private owner and the public authority. Any owner of undeveloped land can submit a plan for subdividing his land into plots to the local authority. The city council can approve such a plan and enter into an agreement with the landowner, provided that the proposed plan is in line with the *PRG* and in the public interest.

The *Programma pluriennale di attuazione* (*PPA*), the multi-year implementation programme was introduced in 1977. All major communes with populations above 10,000 people are required to produce a *PPA*. The scope of the *PPA* regarding general master plans is procedural, as the *PPA* cannot propose alternative plans. Rather, it indicates priorities by introducing the "time factor" into a physical master plan, in which no time limit for proposed changes has been set. In general the *PPA* has three main goals:

- to locate and co-ordinate proposed changes by issuing a time framework for their implementation
- to link urban planning at city level with the economic and financial planning of the commune
- to enforce the implementation of the plan by compulsory development of those areas that should be developed.

In practice, after some initial hopes, the PPA has been little more than an uncritical listing of "intentions to build" as declared periodically by land and property owners. Local authorities' power of enforcement is extremely weak, owing to their lack of financial resources to implement an expropriation for public use. So, the failure to develop a property according to a plan is never fully penalized, as compulsory purchase is extremely unlikely. Nonetheless, the PPA was the first acknowledgement in the Italian urban planning system that time is a factor that matters and should be considered by local policy-makers.

Indeed, the PPA could also play an important role in reformed Italian planning legislation. The experience gained in the past two decades shows that the communes should be encouraged to see the multi-year implementation programme as a strategic planning tool and not as additional paperwork. To achieve this, the PPA should be streamlined and made flexible. For example, it would be a good idea to include in the programme only the "intentions to build" that are above a given quantitative threshold; that is, only the projects that would have an impact at least at neighbourhood level. At present there are no distinctions between, say, a private owner willing to build an underground garage and a real estate promoter seeking to develop a 300-room hotel.

The time of the PPA could be kept at its present length of three years, but the enlisting procedure should be made more flexible so that new entries could be made periodically (e.g. every four months). Finally, the communes should recognize that private developers who complete proposed projects within the stated time limits ought to receive some sort of acknowledgment from the city. For example, the mayor could enrol those firms in a special roster of firms whose future planning applications would be processed (independently from the positive or negative response) in half the time of regular applications. It would cost almost nothing to the commune, and it would certainly be a valued incentive for the developers. Conversely, the commune should also be put in a position to sanction the behaviour of those developers who do not implement the proposed projects, as this may weaken the planning provisions of the commune and generate costs for the urban infrastructures unnecessarily built in the meantime.

A recent challenge to local governments is given by the need to relate urban planning with the planning of financial resources at city level. Since 1994 the budget of many communes has been greatly influenced by the revenues collected through the ICI (the local tax on real estate assets), which have been entirely allocated to the communes in substitution for an equivalent amount of financial resources allocated by the central state. At present, the PPA gives only an estimate of the revenues that would be paid to the commune as *oneri di urbanizzazione*, that is, as "development tax" related to the cost of provision of public services associated with the proposed new constructions.

It was the Law of 1977 (known as *Legge sui suoli*, the law on urban land) that introduced the important principle of *oneri di urbanizzazione*, as described below, and the concept of "public concession" applied to the private development of land. Since 1977, planning permission has become a "concession" granted by the local authority upon payment of a certain sum. The amount to be paid is in two parts, which are proportional to the "urbanization" costs and to the building costs related to a specific project. The rationale behind the first part is that most building and urban changes entail costs for the city; thus, the holder of the building concession has to pay a sum proportional to these costs. The second part is justified on the grounds that the building process is a "concession" by the local government; thus, it requires a payment separate from the amount due for "urbanization" costs. In practice, the costs for planning permits are calculated through standard tables, which are revised periodically according to regulations set out by the region and by the commune.

As the "development tax" is still a cornerstone of the development process, it seems useful to describe in some detail how it is calculated. The tax is split into two levels: for primary public services (sewage, roads, etc.) and for secondary public services (primary schools, green areas and sports facilities). It should be noted that communes can impose a range of development tax rates according to the type of new property developments. The tax for primary public services is calculated by setting a figure in lire per m^3 of construction and a second figure per square metre of land plot. These figures need not to be the same for all kinds of buildings and, in practice, are usually higher for luxurious residential buildings (being identified as "luxurious" according to some quantitative parameters set by law), average for regular housing, and lower than average for condominiums with flats not greater than $130 m^2$ and for certain types of cooperative housing. Likewise, development taxes can vary between shopping centres (usually taxed at the highest rates), industry and small and medium-size enterprises. The lowest tax rate can be, in practice, as low as 60% of the highest tax rate.

In the case of the development tax for secondary public services, the tax is linked to the estimated cost imposed upon the commune by each new resident (usually estimated as a ratio of one new person every $100 m^3$ of new residential building), or by each new job place in new industrial, retail or office developments (here the ratio is usually one new job per $100 m^2$ of floorspace). As in the case of primary services, the commune can impose (and this is done frequently) a different tax weight for each of the various types of property developments. For both primary and secondary public services, the tax rates are left to be determined on a case-by-case basis with regard to certain types of property development, such as low-cost housing, certain types of cooperatives, public buildings, or buildings of public interest.

Thus, the *oneri di urbanizzazione* (development tax), is an interesting tool to influence the property market according to the planning goals of a commune. The tax could be set at more favourable rates for the kind of development considered most necessary in a certain area of the city. Conversely, the development tax rates could be set at high levels for the type of development that should be discouraged according to the stated objectives of a city master plan. In practice, the development tax is a weak instrument for three reasons: first, it can be applied only to new developments, while the largest Italian communes do not have any more expansion areas within their boundaries; secondly, the development tax is simply taken as an additional cost of construction and is passed on to the end-users; thirdly, and most importantly, the absolute level of the tax and the range of variation between property types are limited and do not seem large enough to influence the location decisions of property developers. After some initial conflicts, the payment due to the local authorities to obtain a planning permit has been accepted by real estate developers as just one additional cost to be included in the financial appraisal of any project.

Although the development tax is basically a lump sum paid once and for all in order to obtain planning permission for new projects, the ICI (the city tax on real estate properties), is a recurrent tax on new and old real estate assets. The PPA could be enhanced as a planning tool by introducing an estimate of the various financial revenues generated on a yearly basis for the benefit of the commune budget. For each new building, the concerned commune will increase its annual budget by a certain amount of lire paid as ICI. Other resources will also be generated, for example the tax for the removal of solid waste. Indeed, it would seem a good idea to add to the information included in the PPA the estimates of the various taxes generated, first of all the ICI. The commune should be given the opportunity to differentiate the ICI according to the different types of buildings that fit best with its urban planning goals. In a city flooded with vacant offices, for example, new office projects might be discouraged by a high ICI rate, and traditional retail properties could be given a tax break if a commune decided to expand its supply of tourism services and counteract the decline of viability of central areas caused by large shopping centres located outside its jurisdiction.

Land expropriation for public use

The expropriation process in Italy is commonly regarded as an extremely difficult subject, especially for foreigners. In truth, it is a complex issue.

The most conventional way to explain the expropriation system in Italy is to present, in sequence, all the many laws that affect this domain, starting with the Law of 1865 which remains the fundamental law for land and real estate expropriation. It stated that compensation for the confiscated property had to be at "the right price that, according to technical consultants, the property would have had as the result of a free trade". This amount had to be discounted by the benefits that the construction of a public work on the taken property would have created on the part of the property not expropriated. These principles were amended in 1885 when it was established that the settlement amount should be equal to the actual (current use) value of the property plus the sum of the rents in the previous ten years. This new criterion allowed much smaller compensation amounts; thus, it has been systematically used by public authorities in Italy and repeated in subsequent laws: in 1907 for railways constructions, in 1927 for construction of airports, in 1938 for the implementation of planning schemes of many Italian cities, in 1949 for INA-CASA, the national programme of social housing in the post-war Italy, and in 1955 for the construction of roads and motorways. In 1970, with the institution of the local government bodies, "regions" (*regioni*), the expropriating powers were transferred to the local levels of government: regions and communes. There was a need for clear rules for the calculation of the settlement amount. This was the purpose of the Laws of 1971 and 1977. Both laws introduced in a clear-cut way the distinction between "developable land" and "non-developable land", basically identifying the former with urban areas and the latter with agricultural areas.

This notion was not entirely new. In fact in the expropriation procedure of the Law of 1885 (which is still in effect) the distinction between the two types of land (developable and non-developable land, which includes agricultural land) was already accepted. The novelty of the 1971 law (and its modifications) is the differentiation in the amount of compensation for the two types of land. In the case of agricultural land (i.e. land outside the city), the compensation amount was set equal to the average value of agricultural land for the actual type of agricultural product on the land to be purchased. In case of developable land (that is, land inside the city), the compensation was equal to the average agricultural land value for the most rentable agricultural product in the area, multiplied by a factor of between 2 and 5, for communes up to 100000 inhabitants, and between 4 and 10 for the cities above this threshold. However, this criterion was ruled non-constitutional by the Constitutional Court (*Corte Costituzionale*), in 1980 and again in 1983 when the court also ruled non-constitutional the compensation amounts proposed by the 1980 law. (The latter had basically reintroduced the same criteria as the 1971 and 1977 laws.)

There was a legislative vacuum until 1990 when it was established that all the expropriations to be carried out for *Roma Capitale* (a special programme of public works to upgrade the image of Rome as the capital of Italy) were to be conducted according to the Law of 1885, until a new general law was passed. This "general law" has not yet been enacted, but in 1992 a new law was approved. This has set new criteria for the calculation of the reimbursement amounts due in case of public expropriation of private land and real estate properties. Although it is not yet regarded as a "general law", it is considered to be progress, since it at least clarifies the ways to calculate the settlements and the procedural phases of the expropriation.

The criteria to be followed in order to classify a plot of land in one category or another was set out in a decree by the Ministry of Public Works in 1988. Basically, a land plot is considered "developable" (for calculating the expropriating compensation) whenever a planning permission could be granted, according to an existing planning scheme, at the moment that the expropriating procedure begins. In practice, all land plots within the urban area (whose limits have been defined by the law of 1967) are considered "developable", as are the areas that have a "clear vocation for development", although they are located outside the urban areas. The "clear vocation for development" is defined, according to the Upper Court (*Corte di Cassazione*) by the public authority taking due regard to a set of relevant factors (e.g. the accessibility of the area, the existence of communications networks, the development of nearby areas, and the presence of public services, which require a local community). These criteria will be used in the creation of a new decree by the Ministry of Public Works, which will state comprehensively which lands are "developable" and which are "non-developable". Although at the time of going to press the decree has not yet been issued, it is generally expected that it will reaffirm the rules described above.

The process of expropriation

The listing in chronological order of all the many laws that affect expropriation is useful, because it gives an idea of how the issue of land and real estate expropriation has always been a focal point for Italian society in general and not solely for the planning profession. For example, one cannot forget that the great Italian novelist Giovanni Verga wrote some of his masterpieces (*Vita dei campi, I Malavoglia, Dal tuo al mio*) on the subject of personal belongings, including rural land, in the late nineteenth century. The movie *Le mani sulla città* (*Hands on the city*), directed by Francesco Rosi and shot in 1963, is an unforgettable portrait of real estate speculation in post-war Naples and, for that matter, in the whole of Italy.

In the political arena the *SIFAR* affair, the aborted tentative coup d'état in 1964, was a reaction by the most conservative forces against the extension to the land sector of the recently announced nationalization programme of the first centre-left government of the post-war Republic. That government successfully carried out the nationalization of the electricity sector and the founding of *ENEL*, the state company for electricity, which is now likely to be privatized (in a general context completely different from that which justified its nationalization in the 1960s). Democracy in Italy was not in real peril because of the *SIFAR* affair. However, what cannot be denied is that the event caused the Christian Democrats (*Democrazia Cristiana*), who were the majority party of the coalition in power at the time, to announce on television their withdrawal of support for their own Minister of Public Works. He had gone so far as to work out in detail a legislative proposition that would have separated the right of owning land from the right to develop it. This was clearly too much. Whereas in the case of the nationalization of the electricity sector there were private lobbies that unsuccessfully attempted to halt the project, the private interests in the land sector undoubtedly had much more power and leverage, and the minister left government overnight and played no further role in politics. It must also be said that Italian society was perhaps not ready for such a radical reform. If approved, the separation of the right of owning land from the right to develop the same land would also have been very difficult to implement at the local level, in the absence of a general reform of the planning system and of local finance.

The expropriation procedure

At present the expropriation procedure consists of five phases.

Phase 1: Preliminary phase – document gathering The institution with the power to expropriate (state, region, province, commune, other public institutions, or institutions in charge of works declared to be of public interest) – hereafter indicated as "the expropriating institution" – declares through its legal representative (e.g. in the case of cities it is the mayor; in the case of regions, it is the president of the region) the intention to expropriate a given area. It also indicates the source of financial resources needed for the expropriation and the list of owners (names and addresses, plus their cadastral data) who will face expropriation. The act of expropriation has to be sent to the controlling body of the institution, to be examined for consistency. Having passed this examination, the act receives the endorsement (*nullaosta*) and becomes effective.

Phase 2: Beginning of the expropriation procedure

(a) In the commune where the property is located, the expropriating institution deposits the following documents: a report on the public works for which the expropriation is required, a detailed cadastral plan indicating all lots and a list of owners concerned, and an extract of the *PRG* and other documents not strictly prescribed by law (usually a description of the agricultural uses of the plot and of the buildings that will be expropriated).

(b) The mayor informs the owners of the beginning of the expropriation procedure and publishes it in the *Foglio Annunci Legali* (*FAL*).

(c) Within 15 days after the publication in the *FAL*, the owners can present their observations to the expropriating institution. All the observations must be examined within the following 15 days. The file (which now includes the observations of the owners and the comments of the expropriating institution to these observations) is then transferred to the public body which has the duty to estimate the amount of the provisional compensation (usually the region, although in a few cases the regions delegate the work to the communes).

Phase 3 Estimate of the provisional compensation Within 30 days of the receipt of the file, the region (or the delegated commune) indicates the amount of the provisional compensation, which is calculated with regard to the type of land being expropriated. There are two cases:

- the property is developable land; in this case, the settlement is calculated according to Article 5 of the 1971 law
- the property is agricultural land; in this case, the compensation is calculated according to Article 16 of the same law.

Compensation for expropriation of developable land

The amount of the compensation is calculated according to the following formula:

$$AC = \frac{AV + \Sigma CR \text{ last 10 years}}{2} \times 0.60$$

where:

AC = amount of compensation
AV = actual market value of the property
CR = rent value as given by the cadastre books.

The amount of the compensation (AC) is therefore equal to 60% of the average of the market value (AV) of the land plot being expropriated, plus the sum of the yearly land rent (CR) as it appears on the books of the cadastre in the previous ten years, although updated according to official indices. As this figure (CR) does not change frequently, this summation is very often equal to the CR of the last available year multiplied by 10. Usually, the cadastral rent is a minor figure compared to the market value of the land, and it is also very easily identified through the cadastral office. Hence, the amount of the compensation (AC) is basically influenced by the actual market value (AV) of the property. "Actual market value" means the value of the property at the moment of the official declaration of a public interest by the expropriating institution, that is at the end of phase 1 as described above. According to the Upper Court (*Corte di Cassazione*), the "actual market value" must be estimated according to standard estimating procedures based upon a comparison with similar properties sold in the area.

In conclusion, the settlement for expropriating a privately owned land plot is roughly equal to 50% of its market value, minus a further reduction of 40%; thus the amount of the compensation is about 30% of the market value of the land at the initiation of the expropriating procedure. At present, the amount of compensation is then reduced further by about 20%, because this payment is treated as "earnings" and, according to the 1991 Law, the expropriating institution must apply the urban area tax to this amount at source. In the end, what goes into the pockets of the landowner is as little as 20–24% of the market value of its property. Nevertheless, the Constitutional Court (*Corte Costituzionale*), has so far consistently rejected all allegations of violation of Article 42 of the Constitution, which requires that the settlement offered to expropriated owners of private land must not be "symbolic" but linked to the actual value of the assets. In fact, the Court has always maintained that Article 42 does not call for a complete restoration for the loss of properties; to be constitutional the reimbursement needs only to be "not symbolic and abstract" but linked to (although not equal to) the actual value of the land. This level of compensation is indicated by the Constitutional Court as the maximum that the public institutions can afford to pay to the expropriated private owners in order to implement projects that benefit the whole of society.

Having been informed of the amount of provisional compensation, the owner has three choices:

- voluntary sale of the property
- acceptance of the proposed compensation
- non-acceptance of the proposed compensation.

Voluntary sale of the property Throughout the expropriating procedure the owner can voluntarily sell the property to the public authority, thus avoiding all the phases of the expropriation procedure. For doing so, the owner receives from the expropriating institution a "premium" over and above the settlement calculated above, in order to recognize the time saved by the public institution in obtaining the land and thus in implementing the project for which the land plot is an integral component.

In case of developable land, the compensation is not reduced by 40%, and is further increased by 50%. In practice:

$$\mathrm{ACv} = \frac{\mathrm{AV} + \Sigma\mathrm{CR\ last\ 10\ years}}{2} \times 1.50$$

where:

ACv = amount of compensation for voluntary give away of the land

AV = actual market value of the property

CR = rent value as given by the cadastre books.

In cases where the owner is a peasant and the property is agricultural land, the settlement is multiplied by 3. The amount of the settlement (before being multiplied by 3) is equal to the value of agricultural land in relation to the products actually grown on the land plot. The increment of 50% does not apply in this case.

Acceptance of the proposed compensation Within 30 days of the notification of the proposed provisional compensation, the owner can accept it through a formal communication to the region and to the expropriating institution, if different from the former. In practice, this is only a theoretical event, because by doing so the owner is not entitled to the abolishment of the 40% reduction, and the payment is not due until the end of the expropriation procedure, which is fairly long, even in cases of non-opposition from the owner.

Non-acceptance of the proposed compensation Sometimes the owner refuses to accept the proposed settlement from the public authority, or does not answer at all, which is always taken as a refusal. The expropriating institution will then ask the region (or in some cases the commune) to deposit the proposed compensation amount in a special bank account held at the *Cassa Depositi e Prestiti*, the operational branch of the Treasury.

Phase 4: estimate of the final compensation Within 30 days from the notification of the proposed provisional compensation, if the owner refuses it, the president of the region or the mayor of the commune, if so delegated, will ask a special provincial commission to calculate the final compensation amount. This commission must respond within 30 days. The notification of the final amount is then published on the *Foglio Annunci Legali* (FAL) and from that moment the owner has 30 days to "voluntarily" hand over the property to the public authority, thus avoiding the reduction of 40% of the calculated compensation amount.

Phase 5: enactment of the expropriation act The expropriating institution must provide evidence that the owner has accepted the proposed compensation or that, in case of denial, the corresponding amount has been deposited for this purpose at the *Cassa Depositi e Prestiti*. Within 15 days, the region, or the commune, if so delegated, does the following: it informs the owner, publishes the expropriation act on the *Foglio Annunci Legali* (FAL), and on the official bulletin of the region, transcribes the act at the office of real estate deeds (*Conservatoria dei Registri Immobiliari*), and at the cadastre (*Ufficio del Catasto*). With these bureaucratic acts the expropriation procedure is completed.

The complexity of this procedure is plain. This is why, in all phases, the law of 1992 allows the owner to obtain a premium (the forfeit of the cut of 40% of the calculated compensation amount) if he agrees to hand over his property "voluntarily". Such sales by-pass many phases of the procedure and saves time and, more importantly perhaps for the local government, it permits utilization of funds available to the local authority when the expropriation procedure begins but which may not be available later if the procedure is left to its own pace and timing.

Compensation for expropriation of agricultural land

Compared to the case of developable land, compensation for expropriation of agricultural land is a rather easy case. The compensation is calculated according to the Law of 1971. The compensation amount is set equal to the value of agricultural land in relation to the products actually grown on the land plot in question. In case the owner is a peasant who personally cultivates the land, the settlement is multiplied by 3, if he accepts "voluntarily" to hand over the property to the expropriating institution, as in the case of developable land. The purpose of this premium is to offer an incentive to the owner in order to save time for the community. The law also applies in the cases where the land to be taken is neither developable nor agricultural; for instance, abandoned areas or unused quarries.

In the 1990s some planners have often referred to the so-called "Cutrera Law", omitting to clarify, especially to the foreign observers, that this was nothing more than a proposal by Senator Cutrera, and that it was never passed by the two houses of parliament. According to this proposal, all landowners would receive a minimum "planning permit" of 1 m^3 of construction per m^2 of land. The development rights above this limit would be granted by the local authority only through payment of a substantial development tax, provided, of course, that the application would meet all the planning requirements in the area. Taken to the extreme, this proposition was a new version of the Sullo's idea of 1964, separating the ownership of land from the development rights, only giving to landlords a uniform and limited threshold of development rights. It seems unlikely that this proposition will ever become law, but the issue of development rights is surely crucial for any modern society and the Cutrera proposal contributed to the debate on this subject.

Information systems

The main source of information for the land and real estate sector is the Cadastre, which is a public office within the organization of the Ministry of Finance (*Ministero delle Finanze*). It is composed of two independent sections: the land cadastre (*Catasto Fondiario*) and the urban real estate cadastre (*Nuovo Catasto Edilizio Urbano*, NCEU). The land cadastre contains the information on rural land and rural real estates. The elementary unit is the "cadastral land parcel" (*particella catastale*), a single piece of land, located in one single commune, owned by one owner, classified in one class, and having one utilization. The urban cadastre gives the same information on urban real estate properties and on urban land. Here, the elementary unit is the *unità immobiliare*, the smallest immovable property that can independently generate revenues (a flat, an office, a garage, etc.). In both the land and the urban cadastres, the information is organized in plans, maps and registers. Each property unit is identified by a series of letters and numbers which constitute the cadastral identification codes. The purposes of these cadastres are to:

- ascertain the ownership for each single land property parcel or real estate unit
- classify each property within each commune in cadastral categories and classes
- determine for each property unit a "cadastral rent", which is the basis for the taxation of real estates properties in Italy.

The cadastral rent varies according to location, size, category and class of each property. In practice, however, the cadastral rent depends also on the year of registration at the cadastre. Generally speaking, newly registered properties tend to be attributed a higher cadastral rent than comparable properties registered several decades ago. The kind and amount of cadastral rent is determined for each property by the *Ufficio Tecnico Erariale* (*UTE*), which is also an office of the Ministry of Finance. Both the cadastre (rural and urban) and the *UTE* are organized on a provincial basis, and cover the whole of the national territory.

In the past few years, two important moves have been made by the public authorities in order to enhance the efficiency of the cadastres as instruments for the determination of the identity of the owners of real estate properties and, ultimately, as instruments for tax collection from the land and property sectors. A recent decree by the Ministry of Finance has imposed a revision of the allocation of real estates properties in the land cadastre. As a result, to be included in this register a building must be permanently occupied by a farmer or be utilized by a farmer for agricultural purposes. The decree has also clarified in detail the criteria used to define a "farmer": in brief, this is a person who provides at least one-third of the workforce on a farm, or who gives at least two-thirds of his time to the farm or who derives at least two-thirds of his personal income from agricultural work. In case the owner does not meet these requirements, the building is automatically registered in the *NCEU* (the urban cadastre) and, as a consequence, is assigned to a category and class that entails a higher "cadastral rent". The end result is that the owner has to pay higher personal income taxes and also higher transfer taxes when the property is sold or inherited. The decree was prompted by the fact that in several regions (e.g. around Siena and Florence in Tuscany, on the hills of the Langhe in Piedmont, or on the mountains facing the Riviera in Liguria) many rural houses have been bought by non-residents, often from foreign countries (mainly the UK, Germany and the USA) and subsequently they have been completely renovated. The new houses, albeit fashionable villas, were still registered as rural houses, subject to little or no taxes.

A second provision to curtail fiscal elusion is the law on the so-called "electricity cadastre". The idea was simple. As every house needs electricity, the national electricity company (*ENEL*) is required to indicate in all contracts the fiscal code number of the individual or company responsible for the supply, plus the cadastral identification codes of the property. This enables a quick and easy cross-check of the electricity contracts and the income tax forms of the property owners. These measures have been applied in a less than perfect way, as the efficiency and the productivity of the public administration varies considerably across the national territory. Nonetheless, one can say that they are a success and they have contributed to a significant increase in the tax revenues from the land and urban property sector. It has been estimated that the tax burden on the housing sector alone has been multiplied by a factor of five between 1980 and 1995. Despite this the taxes levied on an average standard apartment of a market value of 300 million lire are still much lower than the global taxes (value added tax, special tax on gasoline, tax on vehicle ownership, etc.) which are imposed on a medium-size car worth about 30 million lire. Compared to the car sector, it is not an exaggeration to suggest that, until the early 1980s, owning real estate properties in Italy was basically tax free.

Since the mid-1970s the wealth of Italian households has shifted from revenues generated by the application of labour to revenues spawned by accumulated assets. The ageing process makes the elderly a section of the population with social and economic characteristics quite different from those in the past. Today a growing number of elderly people do not fit the traditional image of persons with declining wealth resulting from being out of the labour force. Italian pensioners flock to the most popular tourist resorts in Europe and beyond, like their counterparts in Japan, Germany and North America. The tax system should perhaps be reformed to take major social changes into account. For example, the overall burden of taxes on labour revenues could be linked to the age of the taxpayer and be related more strictly to marital status, so as to become lighter at the beginning of a career and for young couples with small children.

3.2 The financial environment

General observations

The framework of the Italian financial system, its organization, functioning and current financial practices, make it particularly difficult to draw a clear distinction between the financing of land and property in urban areas. The financing of urban land by financial institutions is not currently considered to be an attractive proposition for the lenders, neither does it fall under any well defined regulatory framework associated with the financing of future building work. This perceived lack of appeal is explained below.

- Investments in land are vulnerable to changes in national laws which determine the regulations affecting land and the way local government bodies issue planning permissions. The uncertainties over planning permission on urban land tend to discourage financiers, because of different and frequently unpredictable political and social pressures.
- Banks generally regard the financing of a land purchase, and construction work upon it, as good business, because the likelihood of making a profit is high and the possibility of loss is consequently reduced. From this standpoint it is important to emphasize that a bank does not consider someone applying

for a loan solely for land purchase to be really credit-worthy. More likely, this would be interpreted as an indication of inadequate personal resources and a probable inability to finance any future development programme.

- The Italian regulatory framework for real estate financing does not allow for the financing of urban land alone. Any loan granted by a finance house under these rules must be connected with the purchase of existing buildings (*credito fondiario*) or with building construction or improvement (*credito edilizio*). In the latter case, a loan may be granted for land purchase alone, provided that planning permission is already granted, in order to demonstrate that the loan will be used for building work. Outside this regulatory framework, commercial banks can finance urban land investments, like any other type of business, but for the first and second above-mentioned points they do not usually do so except for "very important customers", and then only with short-term agreements typically secured by a mortgage on the land.

The Italian financial system and the financial framework for real estate financing

Within the Italian financial system, the financing of real estate depends primarily on the banks and finance houses, and their system of long-term credit financing, *Istituti di Credito Fondiario* (ICFs). The structure of the financial system is outlined in the following scheme:

Government bonds market

Direct markets* – Private bonds market
– Stock market

Indirect markets

Types of financial intermediaries

- Banks (commercial banks, saving institutions, credit cooperatives, rural banks)
- Short-term financial intermediation

Credit system

Long-term credit institutions

- industrial credit institutions
- agricultural credit institutions
- real estate credit institutions

Long-term financial intermediation

Other credit institutions – Leasing companies
– Factoring companies

Institutional investors – Insurance companies
– Mutual funds (investment/unit trusts)

* *Note:* The difference between direct and indirect markets is based on the existence of brokers who can direct or allocate savings. In direct markets, savers directly purchase the deficit unit liabilities (e.g. stocks and bonds issued by corporations, securities issued by government), whereas in the indirect markets the financing takes place through a financial intermediary. In the indirect market, savers finance the intermediaries by purchasing their liabilities (bank deposits, mutual funds' certificates, etc.) and the intermediaries finance deficit units (business firms, individuals, local authorities, etc.) by granting credits or purchasing their securities.

The current structure of the Italian financial system is based on the "specialization principle", under which the different areas of financial services are managed by specialized institutions. As the above suggests, the specialization principle is applied both within the standard credit system, where a distinction is drawn between the short and long term, and separately from it, where leasing and equipment financing are offered by specialized institutions, and securities management is provided by institutional investors.

The "separation principle" originated in the 1930s, when the most important financial institutions of that time, who were then entitled to enter into any field of financial activity, collapsed and were saved by government intervention. At present, the most important banks are still state-owned, but a process of progressive privatization has been approved and is being implemented (e.g. the state-owned bank *COMIT* was sold to private investors and the general public in 1994).

The second outcome of the 1930s banking crisis is represented by the approval in 1936 of a Banking Act that still controls the organization and functioning of the Italian financial system. The most important principle stated in the Banking Act is the differentiation between short- and long-term credit institutions. The former are banks in their purest meaning, which are allowed to mobilize savings through demand and short-term deposits, and to invest these funds by granting typically short-term loans. The latter are entitled to use savings by issuing long-term bonds or medium-term deposit certificates, and to invest these funds by granting medium- and long-term loans.

The designation of banks and finance firms as short-, medium- and long-term institutions is determined by the regulations:

- short-term financing does not exceed 18 months
- medium-term finance ranges from 18 months to five years
- long-term finance extends for more than five years.

In recent years, a partial deregulation of this separation principle has been introduced. Financial authorities have given banks the option of offering loans exceeding short-term maturities up to a given limit. This limit relates to domestic currency loans and it varies with the ratio of equity and deposits, according to Table 3.1.

Besides the "time separation principle", the organization of the credit system is characterized by a "functional separation principle". The term refers to the

Table 3.1 Bank loans and equity rates.

Deposits: equity ratio as % of deposits	Maximum amount of loans above 18 months
up to 8	6%
8–10	7%
10–12	8%
12–15	9%
above 15	10%

Note: these limits may be increased by an additional 5% ratio if the bank also raises funds with medium- to long-term maturities.

fact, particularly inside long-term credit institutions, that the intermediaries specialize in financing pre-determined groups of borrowers and economic sectors. In practice, long-term credit institutions are classified into three categories: industrial, agricultural and real estate.

Outside the credit system the specialization principle is respected, because banks are not allowed to enter into other fields of financial activity, such as leasing, factoring, investment in securities, and particularly in investment in stocks of industrial corporations. Many banks have at this time been able to overcome these limits, by sponsoring the creation of long-term credit institutions, leasing and factoring companies, mutual funds and, from 1991 on, securities houses. At present, a large part of non-banking financial intermediation is owned by banks. In the near future, as a result of progressive deregulation of the banking business, banks will probably be entitled to promote insurance companies as well as to purchase (to a limited extent) stocks issued by industrial corporations.

The Italian financial system is predominantly characterized by credit institutions, who are the key actors in the process of financial intermediation. The largest part of personal savings flow to the banking system and to government securities markets and, consequently, the financing of private sector borrowing is primarily dependent on bank lending.

The role played in the mobilization and allocation of financial resources by stock markets and institutional investors is therefore less relevant than the one played by the banking system.Table 3.2 highlights the apparent total absence of stocks from the portfolio of Italian families. However, it should be noted that there are legal ways of owning stocks without disclosure of personal identity, through the payment of a fixed-rate tax on capital gains. In addition, most Italian households entered the stock market in the second part of the 1980s and the phenomenon has reached a certain momentum in the mid-1990s, when hundreds of thousands of Italians purchased stocks of the privatized state companies, including *Credito Italiano* (a major national bank), *INA* (the largest insurance company), *ENI* (oil and chemical), to name but a few. Overall, however, it is true that in Italy personal savings are only marginally invested in company stocks when compared to other major countries.

In addition, the table shows that the financing of long-term credit institutions

Table 3.2 Financial assets owned by Italian families (as of 1988).

Money and short-term bank deposits	27.2%
Treasury bills	24.0%
Medium-term deposits	7.7%
Government bonds	30.9%
Private bonds	6.2%
Stocks	–
Other financial assets	4.0%
Total	100.0%

Source: Bank of Italy.

is not adequately provided for by direct personal savings investments, because of the strong competition from government bonds in the pooling of medium- and long-term resources. As a consequence, an important source of finance for long-term credit is obtained by placing their bond issues with the promoting banks or by starting credit terms with them.

Real estate finance: institutions and regulations

Real estate finance is based on two fundamental institutions: banks and specialized real estate long-term credit institutions, *istituti di credito fondiario* (ICFs).

Since real estate financing is typically long term, the possibilities for Italian banks to operate in this sector are limited. The most important field of banking operations in real estate is thus represented by the short-term credit facilities granted to construction companies and the like. Throughout the 1980s at least, the number of mortgages granted to individuals for buying or building a house rapidly increased. This trend occurred because banks considered mortgages to be particularly attractive in terms of the interest rates charged, which were usually higher than business loans. A second reason is that there was actually too much competition among the banks for the supply of loans to businesses, so they decided to extend their range of borrowers and services. From the late 1980s the growth rate of mortgages granted by banks to individuals had reached 25–30% per year, much higher than any other credit sector. However, the trend declined sharply in the first half of the 1990s, in parallel with the economic recession and the slowing down of the real estate market.

The 21 existing ICFs are now long-established institutions, and their original Special Ruling of 1866 is much older than the Banking Act itself. These institutions are now regulated by a 1977 law, the basic principles of which are outlined below.

- Separation between:
  - *credito fondiario*: long-term loans granted for purchase of existing buildings (or in order to raise funds by mortgaging existing real estate)
  - *credito edilizio*: long-term loans granted for the financing of construction work.
- A strict relationship between the maximum amount that can be loaned and the estimated value of the real estate investment:
  - *credito fondiario* – the maximum amount available is 50% of estimated value of the property (estimates checked by the banks and ICFs)
  - *credito edilizio* – the maximum amount available is up to 75% of estimated costs, granted on instalments as the construction proceeds.
- Types of allowable real estate loans:
  - Long-term loans are secured by a first mortgage on the real estate with maturities ranging from five years up to:
  - 25 years for *credito fondiario*
  - 35 years for *credito edilizio*.
  - Medium-term loans, also secured by a first mortgage on property, with maturities typically ranging from three up to five years.

All the above-mentioned rules also apply to banks when they deal with real estate financing.There is a principle that the loans granted must match the bonds issued to finance the *ICF*s. According to this principle, long-term loans must be funded in equal amounts and maturities by bonds issued by the *ICF*, and the medium-term facilities must be funded by issues of medium-term deposit certificates.

Since 1978 the Bank of Italy has authorized the *ICF*s to obtain foreign currency loans from foreign financial institutions, guaranteed by the Italian government.

Statistics on real estate financing in Italy are scarce. A survey carried out at the beginning of the 1980s provided the data shown in the Table 3.3. The figures demonstrate the still predominant role of *ICF*s in financing real estate, and particularly in extending mortgage loans to families, whereas banks appear more involved with the financing of construction companies. However, some estimates made in the 1980s show the opposite trend. Throughout the decade the banks increasingly lent to individuals, while *ICF*s increased their lending to construction firms, particularly the firms involved with larger projects (see Table 3.4).

Table 3.3 Real estate financing from credit institutions (1982).

ICFS	
loans to individuals	54.6%
loans to construction firms	7.4%
Banks	
loans to individuals	13.2%
loans to construction firms	24.8%
Total	100.0%

Source: survey made by Catholic University, Milan.

Table 3.4 Loans granted by long-term real estate credit institutions (stocks, billions of lire at the end of 1989).

Mortgage loans to individuals for house investments	56416
Loans to construction firms	8149

Source: Bank of Italy.

The activities of *ICF*s will be significantly affected by changes in the regulations that will occur after the approval by Parliament of new laws. The regulations will potentially lead to the privatization of the state-owned *ICF*s, and to an increase in the maximum amount that can be loaned for real estate investment. Loan maturity dates may not be correlated with investments, and the funding policies of the *ICF*s will generally be more flexible.

Apart from banks and *ICF*s, a relatively recent addition to real estate investment are specialized leasing companies. They lease buildings, typically industrial/commercial buildings, to firms that pay the lease for a relatively long time, say 15 to 20 years. At the end of this period the lessee will have the opportunity to acquire legal ownership of the buildings, by paying a redemption fee to the leasing company.

There are two basic advantages of leasing:

- the periodical rentals are tax deductible as costs and, since these amounts are higher than mortgage interest and the depreciation on purchased properties, there is a higher tax relief

- the investment is fully financed by the leasing company, and not partially as for real estate loans.

The basic disadvantage is that leasing is usually more expensive than ordinary borrowing from banks and *ICF*s.

At the beginning of 1988, according to available statistics, the number of real estate leasing contracts was 2700 and the relative value equalled 1450 billion lire. The distribution of leasing operations per financed sector was as follows:

- 30% industrial properties
- 28% commercial properties
- 34% office properties
- 7% miscellaneous.

A new alternative for the financing of real estate investments may be provided by real estate investment trusts (*REIT*s). These institutions, well known in Germany, Switzerland and France, are not yet operating in Italy, for the want of appropriate regulations. A law introducing and regulating them has been proposed in Parliament, but it has been waiting for approval since the early 1980s.

In order to provide a complete overview of the financial environment for real estate investments, it is necessary to mention the lending activities of private finance corporations and even individuals. These grant credit, particularly mortgages, for the purchase and construction of buildings. Their activities do not fall under any specific regulatory framework, and no statistics are available for them. These forms of financing are often more flexible than those offered by banks, particularly in terms of the amount that can be loaned, the time period and repayment schedules. Also, the time required in order to complete a credit analysis and to grant the loan is usually shorter than that taken by the banking system. The key problem is represented by the fact that the cost of these loans is higher than those for banks and *ICF*s loans. In some cases the interest rates charged are very high, particularly when these companies accept borrowers rejected by the banks and *ICF*s because of their low credit worthiness.

Transaction costs in the land and property markets

Types of financing: the urban land market The above outlined framework for real estate finance clearly suggests that there is no specific financial environment for the urban land markets. Moreover, the regulatory patterns and the consequent structure of intermediaries imply that there are substantial difficulties in arranging finance for land investments. The key regulation of 1977 for real estate loans defines only two types of investments eligible for such loans:

- financing of existing buildings
- financing of new constructions.

The financing of urban land investments is thus excluded from any such regulatory framework, and consequently from *ICF* interventions. As already mentioned, the only case for *ICF* financing on urban land occurs in the period immediately preceding construction and must be supported by the presentation of the planning permission given by local authorities.

The only opportunity for financing urban land purchases is thus represented by ordinary bank loans, which do not fall under any specific regulation, except for the general one on bank lending. For the reasons mentioned above it is difficult to arrange such financing.

In conclusion, there are three basic forms of financing:

- *Short-term unsecured credit line* These loans can be obtained by highly credit-worthy firms, of long standing, and only rarely by other applicants. The interest rate charged is often higher than the average.
- *Medium-term facility secured by first grade mortgage on land.*
- *Long-term mortgage loan secured by first mortgage on land.*

It is important to stress that short-term credit lines often have the nature of "bridging financing", that is, loans granted for a short period, while waiting for the beginning of construction and the subsequent long-term mortgage loan.

In the case of intermediate and long-term mortgage loans for urban land, the most important problem is represented by the fact that the estimated value of the land is uncertain, because of the frequent and unpredictable changes in urban land-use policies. A piece of land eligible for residential building may become eligible for industrial or commercial use or even for no construction at all, and vice versa. The value of the property is the key element in mortgage loans, because it determines the amount of credit that can be granted. Banks are thus led to adopt very cautious and conservative estimates, and grant loans not exceeding 50% of these estimates. The financial support of banks to pure urban land purchases is thus particularly limited.

Transaction costs The basic costs are the following:

- *Intermediation fee* (in case of purchase through the intervention of brokers) Pure urban land intermediation is not the most common activity of real estate brokers. They concentrate their business in properties brokerage. The most common examples of urban land brokerage occur when a piece of land is sold, together with an already approved planning permission for a given type of building. The selling of urban land outside this case falls under the above described precautions and is frequently arranged not by the usual real estate brokerage firms, but by individual brokers, performing a specialist activity. The identification of intermediation fees is thus particularly difficult. In addition to this, the fees vary widely from case to case, according to the cost of the operation, the location of the land and the type of broker. A relatively common standard minimum fee is 2% of the land price. It is known that fees of around 8% (and even more) have been applied in some instances.
- *Costs of financing* Because loans on urban land are not favoured, the interest rates applied tend to be higher than the average. At present (1996), the prime rate in Italy is around 9% and the usual urban property mortgage loans are charged at an average rate of 14%. The few pure urban land loans

arranged are made at rates above this, unless the borrower has a very strong bargaining power. As for brokerage fees, the scarcity of such deals renders the definition of standard conditions particularly difficult. It is assumed that they are arranged case by case.

Other costs connected with the loan The other costs connected with the loan are:

- a fee for the credit analysis procedure and for the land appraisal by the banks experts: 250000 lire minimum
- a tax on long-term loans (it is applied only to loans with maturities exceeding 18 months): 0.25% of loan amount
- a legal fee for writing the mortgage security on the land and for the contractual agreement between bank and borrower: highly variable, according to the type of requirements.

Types of financing: the urban property market The Italian regulations and banking practices only devote attention to the financing of the property market, rather than to the land market. The key financing institutions are consequently the above-mentioned ones: banks, ICFs, real estate leasing companies and generic finance companies. For banks and ICFs, the most important financing institutions, the regulations deserve special attention, as they define different criteria for the financing of existing buildings (*credito fondiario*) and the financing of new constructions (*credito edilizio*), according to the already described patterns. The two basic types of loans secured by first mortgage on property are long-term and medium-term.

Given the evolution in banking practices since the mid-1980s, the usual (current) technical features of the first type of loan are described below.

Maturities Typical maturities are 10–15 or 20 years (less frequently), that is, below the maximum maturity permissible by law. This is because of the uncertainties in assessing the credit-worthiness of the borrower over a very long period of time, the value of the property and the trends in interest rates.

Repayment schedules Repayment schedules are based on biannual instalments, usually of equal amount, comprehensive of reimbursement of capital and interest due on remaining debt, until full amortization of the debt at maturity. Some banks may offer a grace (or pre-amortization) beginning period, during which the borrower pays only the interest due on the debt, without immediately starting to repay portions of the principal amount. Under this solution, relevant financial relief is provided to the borrower, particularly in the crucial initial period. A few banks have tried to start a new amortization plan, based on progressively larger instalments. This solution may be particularly appealing to borrowers starting from a weak financial condition, but with prospects of a gradual increase in income-earning capacity.

The method of determination of the interest rate The method of determination of the interest rate was fixed until the 1970s for long-term loans. The unpredictability in inflation rates since then has forced the majority of the banks to drop fixed-rate loans, in favour of loans with floating rates.

At present, because of the difficulty in foreseeing future interest rate trends, both of these services (fixed and floating rate loans) are offered to borrowers by banks and ICFs. There are two types of floating rate loans:

- adjustable rate loans, where the rate is adjusted at regular or irregular intervals, according to the bank's discretion; the adjustment frequency is usually every two to three years
- indexed rate loans, where the rate is automatically adjusted at each instalment payment (usually biannually) on the basis of the changes occurring in some pre-selected market index (prime rate, yields on Treasury bills, interbank rates, etc.).

Less frequently the indexation refers to consumer price trends and other direct measures of inflation.

The currency of the loan In the late 1980s many banks and ICFs offered their clients foreign currency mortgage loans. The exploitation of the opportunities for foreign currency financing is attributable to the progressive deregulation of financial transactions in foreign currencies that has occurred in recent years. The rapid growth of such loans is the result of the relatively lower interest rates on foreign currencies than on the Italian lira. In 1989, for example, foreign currency loans accounted for almost 50% of the new real estate loans granted throughout the year. The increasing innovation in financial products has recently led to the introduction of multiple currency loans. In this case the borrower has the possibility to choose the currency of reimbursement.

Obviously, the situation changed completely in the autumn of 1992 when the devaluation (up to about 30%), of the lira against the major European currencies led to the exit of Italy from the European Monetary System (EMS). This meant a sharp unexpected rise in the cost of servicing the loans granted to Italian households in foreign currencies, including the ECU. This burden was substantially greater than the original saving of a few percentage points in the interest rates on the loan. In some cases the households were unable to pay back the loan. This was particularly the situation for households who had asked for ECU loans for their residential house in Italy, rather than the many Italian households who bought a vacation house on the Côte d'Azur or elsewhere abroad. The problem of loans issued in ECUs affects some 20000 people, some of whom have recently set up a committee to bring a legal case against both the Italian government and the banks. The former is accused of not having kept the value of the lira within the official floating range of the EMS, and the latter are charged with misinformation, as they apparently encouraged the households to take foreign currency loans without informing them of the related risks. In 1995 *ABI*, the Italian banking association, instructed the banks to reschedule the loans in order to facilitate their payback. However, other measures demanded by the committee (including the payback of the loan at the official ECU/lira exchange rate before devaluation) seem extremely unlikely. It would seem unfair to ask the government or the banks to take responsibility for this financial burden, because it is the result of a free choice, a speculation in financial terms, by the households. If no devaluation occurred, the borrowers in foreign currencies would doubtless have benefited from lower interest rates and, thus, would have received a financial gain which they would certainly have kept for themselves.

In the case of medium-term loans, similar conditions and features are identified, apart from the maturity, which ranges from three to five years. In addition to this, the repayment schedule of the loan is often more flexible than for long-term credit. Nowadays, the percentage of such medium-term credits is low in comparison with long-term loans. This occurs because the high fixed costs in arranging mortgage financing are worth bearing only for long-term credits. In addition, the repayment burden of a medium-term loan may be excessive for the borrower, compared with the repayment schedule of long-term loans.

State subsidized financing for individuals buying their first house

Individuals buying their first house have benefited historically from fiscal and financial advantages. Indeed, since the 1950s a fundamental pillar of the social policy of the Christian Democrats (*Democrazia Cristiana*) has been the encouragement of access to home-ownership. The rationale was that an increasing percentage of home-owners would promote a high level of social stability. The households with a mortgage to pay back every six months would be less inclined to respond to the "call to action" of the leftist parties.

One typical case of state intervention in providing subsidized loans to first-house buyers is represented by the so-called Goria Law (from the promoting minister's name), approved in 1986. The beneficiaries of this law were exclusively employees less than 45 years old, employed for more than two years, with no other house ownership deemed adequate to meet the family needs in the town where they intended to purchase the new house. The loan granted under this law had a maximum maturity of 20 years and a maximum amount equal to the lowest of the following figures: 60 million lire, 75% of the property price or 2.5 times the annual income of the family. The security was a first mortgage on the property, and the interest rate varied according to the income capacity of the family but did not exceed 13%. The repayment schedule was based on the household's income, according to the key principle that the annual instalments could not exceed 20% of the household's total annual income. The loan could be used only to buy residential buildings, of common type, with no special or expensive features, nor those of artistic interest. Once purchased, the house could not be resold within less than 20 years, except in the case of death or unemployment of the borrower. The benefits provided by the Goria Law were designed only for households with an annual income of less than 50 million lire. The special fund constituted for such state activities was soon fully allocated and no refunding is foreseen in the near future.

This is just one example of many minor laws (*leggine*), which have been approved between the early 1950s and the beginning of the 1990s, in order to facilitate not only homeownership, but also (and primarily) the creation of a social consensus in support of the political coalition in power.

Transaction costs As far as urban property is concerned, transaction costs arise throughout the transfer of the ownership and throughout the arrangement of the financing. The first type of cost refers to the taxes applied in case of property transfers and to the fees paid to real estate brokers, if the transfer has been intermediated. The second type of costs refers to any charge arising for obtaining and managing the mortgage loan that finances the investment.

Transfer costs In cases where the property transfer is arranged through real estate brokers, there is an intermediation fee whose amount varies greatly according to the type of property and to the kind of intermediation services. There is an extreme fragmentation of real estate brokers. According to estimates by Gabetti, a leading Italian real estate agency, there are at present in Italy around 15000 real estate brokers. Their activity is typically carried out on a local basis and they commonly have one or very few branches. An important part of real estate intermediation is performed by professionals usually engaged in other activities (lawyers, agronomists, etc.) and even by individuals, on an occasional basis. A relatively new form of intermediation is provided by the real estate Exchange (*Borsa immobiliare)*, which opened in Rome in May 1990 and in Milan in April 1991. Here the system of prices and the setting of the fees are highly visible and the organization of the market is more efficient. The demand and supply of properties is organized on a computerized basis by the real estate brokers that join the system. The intermediation fees are lower than average, around 2% of the price, charged both to the seller and to the buyer. The disadvantage of the real estate Exchange is that agencies tend to bring to that market only the properties that they could not sell through their own channels.

It is important to underline that the property sales arranged through the registered agencies are only a fraction of the total real estate sales. For example, it is estimated by Gabetti that in Milan only 36% of real estate sales are arranged by real estate brokers and that the percentage is lower in smaller centres. There is consequently a domain of intermediation costs that is at present unknown. However, it is believed that the fees charged on property transfers outside such organized markets vary from a basic fee of 2% up to 6%, or even more in some cases, according to the type of building, the price, the location and the type of broker.

Taxes on property transfers For property transfers the seller and the buyer have to face several taxes. The seller has to pay:

- a tax on the increase of the value of the property (*INVIM*), calculated as a percentage rate on the difference between the sale value of the property and its value at the time of purchase. This tax was terminated in 1993 with the introduction of *ICI*, the local tax on real estate properties.
- income taxes.

The sale proceeds are accounted in the yearly income for the application of income taxes if the seller is a firm. When the seller is an individual, income taxes are only charged if fewer than five years have passed from the date of purchase.

The buyer has to pay four different taxes (Table 3.5). In the most common type of transaction, when a private individual purchases a house from another private individual, the global amount of taxes due from the buyer is 10% of the property price. This is the sum of four taxes: 0.40% is a tax to cover the cadastral services (*imposta catastale*); 1.60% is for to cover the cost of mortgages registration (*imposta ipotecaria*), 8.00% is for a tax imposed for the registration of all contracts (*imposta di registro*) in the Register of property contracts (*Conservatoria dei registri immobiliari*) which is the public office where all the property contracts and the property mortgages are kept. The value added tax (*IVA*) is due only when a property is new and the seller is not an individual but a commercial company.

In sum, the cost of transferring a real estate property in Italy is very high, if one considers that to the 10% mentioned above we have to add other costs (e.g. the cost of the notary public services), which are in the range of 1–3% of the price. This high cost is a hindrance to internal migration flows, especially those motivated by the search for better job opportunities. In terms of pure spatial economics, any property transfer tax would inhibit a perfect match between people and locations. High transfer taxes are also cited as one of the causes of the low mobility within the housing stock within Italian cities. The Italian population is ageing quickly. Many old couples continue to live in large flats, bearing high utility costs, because moving to a smaller house would generate soaring taxes. At the same time, and in the same city, young couples squeeze into micro apartments and often postpone the first child for want of space. From a social and economic viewpoint, it seems undesirable to allow the rented housing market to decline completely, since is the only sector capable of ensuring a free flow of households in the country. By the same token, it would be desirable to decrease the transfer taxes to a nominal fee, while increasing the local tax on real estate properties (the *ICI*) in order to induce a more efficient use of these fundamental assets.

Table 3.5 Property transfer taxes (in lire and percentage of selling price).

	Seller	Buyer			
Purchased from: Tax:	*INVIM*	*CATASTALE*	*IPOTECARIA*	*REGISTRO*	*IVA*
Construction firm (non-luxurious houses)	yes	100000	100000	100000	4%
Others					
Other firms	yes	100000	100000	100000	19%
Individuals	yes	100000	100000	100000	19%
Properties of historical or artistic interest	yes	0.40%	1.60%	4%	no
Properties in foreign countries	no	no	no	100000	no
Cooperative houses	yes	100000	100000	4%	no
Others	yes	0.40%	1.60%	8%	no

Costs related to financing

- *Interest rates* The current ordinary rate on mortgage loans, granted for the financing of property purchase, is at least 2–3 percentage points above the current prime rate. The interest rate may be on a floating or fixed basis. Lower interest rates are charged when the loan is granted in a major foreign currency, whose prime rate is usually lower than that of the Italian lira. But loans denominated in foreign currencies are now very rare, because of the risks connected with the fluctuation of exchange rates.

Another example of lower mortgage rates is where subsidized credit is obtained. In such cases part of the loan cost is borne by the state. At present, in Italy it is very difficult to arrange these facilities, because the most important subsidy programmes have not been refunded. Finally, lower interest rates apply when a co-operative is created for the construction of a new block of flats or houses, but the properties can be sold only to the co-operative's members, who must observe certain time limits before reselling the property.

- *Credit analysis fees* The fee charged by the banks and ICFs for credit analysis work ranges between 250000 and several million lire for the largest operations. The cost depends not only on the size of the loans, but also on the type of bank and on the procedure followed for the analysis. The property value estimate can be done by a bank employee or by an external expert chosen by the bank. The latter circumstances ensure a higher fee.
- *Tax on long-term loans* This tax is applied by banks for any loan exceeding 18 months' maturity. The tax rate is 0.25%.
- *Notary public costs* These costs are borne in order to legalize the mortgage security on the property and to prepare the formal contract between the bank and borrower.

Such costs are usually inclusive of the taxes charged on the deal (apart from the long-term loan taxes charged directly by the bank), primarily the register tax and the tax for mortgaging the property. They may also be inclusive of the costs of inspecting the past history of the property in the property contracts registry (*Conservatoria dei registri immobiliari*), in order to guarantee to the bank that the building is effectively owned by the borrower and that no other party can have rights to it.

- *Insurance fee* Banks and ICFs require an insurance against the major risks of building damage or destruction, the cost of which is borne by the borrower. The cost of such an insurance varies from 1000 lire per million value insured up to 2500 and even more, according to the type of insurance company, the types of risks insured, and the degree of protection provided.

3.3 Tax and subsidy environment

Taxes concerning immovable properties

The Italian tradition of taxation on immovable properties is based on some deep-rooted principles, which are worth describing in order to understand the most recent developments.

The Italian fiscal system is strongly centralized. Apart from some minor taxes and fees affecting immovable properties that belong to the lowest tier of local government (the commune), most of the powers and responsibilities in this field are retained by the national government, including decisions over the tax base, the rate structure, the assessment procedures, the tax collection and disputes. Before the tax reform of 1971, communes and provinces had the power to apply supplementary tax rates to the national taxes on incomes from immovable properties. These financial sources were totally centralized by the 1971 reform.

Whether income or wealth (or a mix of the two) should be chosen is still the subject of endless debates on efficiency and equity grounds. But the outcome is that, since the nineteenth century, the Italian fiscal tradition has not considered wealth as a tax base for recurrent direct taxes. Yet, wealth as a tax base is included under the taxation regime for gratuitous and non-gratuitous transfers.

Since wealth and capital are not considered as a base for recurrent taxation, the Italian fiscal tradition has always accepted both the "concept view" of the taxation of global income (as in personal income tax) and an approach that permits different treatments for different types of income (based on equity criteria). Differentiation between capital income and work income has been justified, when recurrent forms of taxation of wealth are not applied. This means establishing some sort of discrimination, in particular between earned and unearned incomes.

The prevailing mechanism of assessing incomes from real estate properties is based on the Cadastre (*Catasto*). Cadastral values are taken as a proxy of presumed rental values. They refer to different categories and uses of land and buildings. They are based on a concept of average and normal income and they are usually updated at intervals of three years. Cadastral updated incomes are the only term of reference for assessing the following: incomes ascribed to ownership of agricultural land, those ascribed to agricultural firms, and those ascribed to owner-occupied houses.

Taxes on immovable property

Taxes on real estate properties were mainly the responsibility of central government until 1993, when the lowest tier, the commune, was given the power to levy the new tax, *ICI* (the local tax on real estate properties). Indeed, the introduction of *ICI* was a real turning point. Few other taxes (personal income tax, tax on corporate gains, value added tax) have resulted in such a great volume of tax revenues.

In 1994 the *ICI* tax revenue amounted to about 14500 billion lire, about one third of the special financial provision which the government approved with the national budget act at the end of 1994. In 1993 the tax was levied by the communes but went to the central budget, whereas since 1994 it is levied and kept almost entirely within the communes.Their yearly budgets have become substantially dependent on the tax. This may sound familiar to readers whose country (like the UK or the USA) has traditionally imposed hefty taxes on real estate properties, but it was a novelty for Italy. For example, in 1994 the tax was about 1500 billion lire for the Commune of Rome, about 800 billion lire for Milan and 550 billion lire for Turin. In 1993 about 7300 billion lire came from the North, 3600 billion lire from the Centre and 3000 billion lire from the South and Islands.

It must be noted that although for the tax payers *ICI* was a truly new and additional tax, for the communes it did not represent any real new source of revenues: transfers from the state budget to the communes have been cut proportionally. This fact has introduced a new component of unfairness in the distribution of tax revenues. The communes with a large and valuable real estate stock (e.g. the rich tourist centres on the Alps or along the upper coast of the Adriatic Sea) can now rely on a true "gold mine", as their city budgets and the public services for the few all-year-round residents are generously sustained by the owners of second homes, who live in the big cities and who occupy the second homes no more than 50 or 60 days per year. In the centres with no tourist properties and low-value housing (such as many villages in the Appennines or in the lower parts of the Alps), the communes cannot rely too much on the *ICI* to provide local services.

Central state taxation on immovable values The Italian national system of taxation of immovable property covers two different kinds of tax base: incomes and transfers.

Taxation of immovable property incomes Incomes from immovable properties are taxed under the ordinary regimes of the personal income tax (*IRPEF*). The personal income tax and the corporate income tax were introduced by the general tax reform of 1971, replacing the previous system of taxes on incomes. Values are assessed according to the system of updated cadastral incomes or, alternatively, actual net incomes, as already described above.

Government policy has traditionally encouraged home-ownership. As a result, by the mid-1990s, over 75% of Italian households own the house where they live and, in addition, there is also widespread ownership of vacation houses. This figure is rather high compared to other major European countries. Owner-occupiers are taxed according to the presumed rental values resulting from the updated cadastral data. At present, owner-occupiers receive hidden benefits because updated cadastral values are considerably lower than market values. Thus, this fiscal treatment is another benefit given, albeit indirectly, to home-owners.

To a greater extent the same situation applies to agricultural ownership and enterprises' cadastral incomes. It is not an exaggeration to say that in Italy rural properties (owned by farmers) and their commercial activities are almost tax free. At present, interest obligations from long-term mortgage loans can be deducted against tax to a rather limited degree (about 7 million lire for the owner-occupied house). It seems that this upper limit should be replaced by a threshold that could be updated from time to time in order to take account of inflation.

Since the mid-1960s, the tourist house sector has increased enormously. Many houses in the mountains, by the sea or by the Italian lakes are not rented, but purchased for use during holiday periods. Opportunities for speculative investments are often greater in these areas than elsewhere. It is debatable whether these so-called "second houses" (as opposed to the "first house" in the city) should be treated differently from other properties. At present, most utilities are supplied to holiday houses at higher costs than those to residential homes. For example, telephone and electricity supplied to tourist houses cannot benefit from social tariffs, which are, by contrast, applied to the vast majority of the residential housing stock. In addition, cadastral incomes are increased by one third in the case of a tourist house. Nonetheless, these measures do not add up to a fiscal treatment significantly different from the ordinary housing stock. The result is that the tourist housing stock, which has swelled in almost all regions, brings about permanent and serious environmental problems, despite the fact that the tourist houses are occupied for only 60 days per year on average.

In Italy, rent control legislation of residential properties is rather severe in municipalities with more than 5000 inhabitants. The parameters used to calculate the "fair rent" are rigid and do not generally reflect market values. This situation created in the 1980s a huge black market for housing. To discourage owners from keeping houses unlet, updated cadastral incomes are multiplied by three, if a house, which is not used by the owner himself or by a member of his family, is left unrented for a period of more than six months. It must be noted, however, that since the introduction of *ICI* in 1993, the attitude of property owners has changed: many of the housing units previously kept empty on purpose (and unlet) have been put back into the rental market to cover the costs of the new tax.

One of the major features of the Italian fiscal tradition is not to have a general tax on wealth or capital. To take wealth or capital elements of incomes into account, rate discrimination in respect of incomes from different sources was introduced. Although many people supported a general property or wealth tax, this rate discrimination survived after the 1971 reform, thanks to the introduction of a local income tax. In practice a flat rate of 16.2% is applied to all incomes as assessed by personal and corporate income taxes, since wages and salaries, independent and professional workers' incomes, and dividend incomes (the latter already being subject to this supplement as a component of corporate income). This tax was meant to be part of a new system of local taxes. Its revenue should have been given to regional government, but since 1971 this highly controversial decision has not been implemented. Thus, central government is the sole beneficiary of this tax.

Tax on selling real estate Taxation on property sold can fall under three different headings:

- tax on the free transfer of wealth
- tax on real estate transfers
- value added tax.

Tax on the free transfer of wealth (inheritance and gift tax) As in many other countries, tax rates are progressive according to the size of the bequest and according to the relationship between the heir (or the beneficiary) and the deceased (or donor). Two problems have recently been solved with respect to this tax. First, a "fiscal drag" effect can occur in inflationary periods, giving rise to large increases in nominal tax burdens. To cope with this problem, the rate structure was changed in the late 1980s. The second problem concerns the assessment of taxable values. Until recently, assessments were based on market capital values. This evaluation procedure led to a great deal of legal argument between taxpayers and the administration. At present, the taxable values of real estate properties are estimated automatically by making reference to the updated cadastral value (in practice, for an urban property the cadastral income figures are multiplied by 100, and for rural properties by 75). The results are not always in line with market values, and in some cases they are rather odd: old residential buildings tend to be underestimated whereas non-residential buildings are overvalued.

Tax on non-gratuitous real estate transfers The evaluation procedure for this tax is the same as the tax on free transfers. Liability lies with the buyer. The normal flat rate is 8%. But there are several special rates. This encourages "camouflage practices" in contracts. There is a general agreement that the rates should be sharply reduced, thus increasing market mobility.

Value added tax The ordinary regime of VAT applies to property sold by firms. The normal rate is 19%. Company building sales are taxed at a 2% reduced rate. In either case the tax on registered property transfers does not apply.

Local government taxation on fixed valuation

- *Tax on the increased capital value of property (INVIM)* This capital gains tax was introduced in the 1971 reform. The commune is the full beneficiary, but all other strategic decisions (administration, evaluation, assessments, changes in the tax base and in the rate structure, collections, etc.) are the responsibility of central government. This tax is non-recurrent for individuals (capital gains are assessed and taxed when due) and recurrent, at ten-year intervals, for corporations, the rate being progressive. Liability lies with the seller or with the heir. In the latter case, to avoid double taxation, a coordinated scheme is provided between this tax and the inheritance tax. Original and final values are estimated according the updated cadastral values, as previously described.The tax *INVIM* was terminated in 1993 with the introduction of *ICI*, the local tax on real estates.
- *Contributions on obtaining building permit* These taxes (known as *oneri di urbanizzazione*) are levied when a building permit is issued by the municipal authority. They are applied under regional regulations and they are related either to the volume or to the area of the building.
- *Refuse removal tax* This tax is levied by the commune, which is also respon-

sible for deciding the base cost of the tax. It is calculated on the basis of the floor area of the property plus a share of gardens and balconies if present. The occupier is liable for payment. In the past this tax was almost symbolic and was far from covering the pure cost of removing the garbage. Since the communes have been given more responsibilities and fewer resources from central government, the tax levied has increased systematically. The trend has also been sustained by the growing concern for the environment, which has pushed local governments to invest in costly technologies.

Adjustments or reform: the main issues

The tax reform of 1971, which affected both direct and indirect taxes, left two major problems unsolved. First, there were taxation powers and responsibility at subnational levels of general government (20 regions, 103 provinces and 8103 communes). The tax reform was based on heavy central responsibility for taxation, for the following reasons:

- problems of vertical fiscal imbalances, because of lack of resources at local level
- problems of horizontal fiscal imbalances, especially as far as the underdeveloped areas were concerned; in general, this trend towards greater centralization of taxation responsibility led to a greater fairness in legal procedures
- technical difficulties in finding an "optimal taxation" for regional governments.

Under this approach, regional and local authorities should be mostly financed by state grants. As in many countries, regional levels of government have been increasingly financed by a huge expansion in central grants. Two major criticisms are levelled at this system:

- the inefficiency generally associated with the unnecessary separation between public expenditure powers and benefits, and the cost involved in obtaining them
- the disincentives in local government accountability.

Secondly, there was the reorganization of property taxes on the basis of greater equity and efficiency. The present system of taxes affecting property is generally considered to be financially unfair and politically inefficient.

In a White Paper published in 1981, entitled *Reorganizing taxation on fixed property*, the Italian Ministry of Finance suggested a joint solution of the two problems, thus stressing the connection between property taxation and local finances. According to the paper's conclusions, local government should have responsibility within the property field. Two alternative solutions were suggested: either a "special" tax or a tax on capital values of property. The latter solution prevailed and in 1993 the *ICI*, (the tax on real estate properties), was introduced.

In Italy the tax system is a very hotly debated subject. The topic is under continuous and fierce discussion. Alternative solutions fall between a set of adjustments to the 1971 tax structure and more radical reforms. Property taxation and devolving financial responsibilities to local government are two of the most recurrent topics, together with the rates structure and the various loopholes in personal income tax.

In 1989 a new tax affecting commercial, industrial and professional activities was introduced (the so-called *ICIAP*, i.e. *Imposta Comunale sull'Industria, Arti e Professioni*). Here the communes are responsible for assessment, collection, and for avoidance prevention.

Tax revenues are shared between the communes (90%) and the provinces (10%). Liability for payment lies with all persons and firms running a commercial, industrial and professional activity, as classified and identified by the value added tax system. The tax base is determined by the surface area of the premises where commercial, industrial and professional activities are run. Thus, the tax liability lies with the occupier of the building. The revenue accrues to the commune where the building is located. Tax rates are set in annual lump sums determined by a matrix built up by ten activity sectors and by seven area classes. This is the crucial element as far as fixed property is concerned.

Besides this, tax liability is corrected to take into account an income element, as determined by personal income tax or by corporation income tax. Tax rates, as determined by the matrix, are halved when the annual income is below 12000 ECU (approx.), and doubled when annual income is above 33000 ECU (approx.). The commune can increase or decrease the lower limit by 50%, and increase or decrease the upper limit by 40%.

Nonetheless the *ICIAP* system has been considered for some years a transitional solution in place of a new system of local taxation, which was expected to be approved in the early 1990s. In reality there has not been a general reform, although some new taxes have been introduced. The main one is the *ICI*, the local tax on real estate properties; the "tax for local services" is still to come. The main features of these two taxes are described below.

Local fixed property tax The *ICI*, was introduced in 1993. It is imposed by the communes on all properties within their administrative territory, with few exceptions (churches, embassies, non-profit institutions, etc.). The rate is chosen every year by the city administration in a range between 0.4% and 0.6% of a conventional value of the property, which is equal to the updated cadastral rent value multiplied by a factor of 100. In case a commune is in a state of recognized budget deficit, the city council is required to impose a special rate of 0.7% of the conventional property value.

Tax for local services This new tax is intended to provide full public services for the enjoyment of property holders. This tax will result from merging some minor local taxes and the tax affecting commercial, industrial and professional activities, which has been described above. The basic principles are just the same. Again, only non-domestic premises will be affected. Liability for payment will lie with occupiers. The commune will be responsible for collection, and will be the beneficiary. Communes will be allowed to alter the rates for various reasons, e.g. the quality and quantity of local services provided, the area and uses of the building, and taxpayers' incomes.

In conclusion, the new taxes are intended to restore responsibilities for taxation to local government, thus promoting financial accountability. Also, the new system is directed towards balancing the local tax burden between owners and occupiers, with special attention being given to owner-occupied houses. Taxes on property includes all fixed properties, irrespective of their domestic or non-domestic use, thus allowing a fair solution to the rather artificial "instrumental goods" description applied to industrial real estate.

Finally, the *ICI* tax is expected to shift the present Italian system of property taxation from an income and transfer tax base to a system based on capital value.

Public subsidies on land and property markets

In Italy, subsidies and incentives exist in the provision of housing. They are intended to benefit particular categories of people (the disabled, pensioners, refugees, immigrants, the military and public dependants). Other subsidies have been created with different aims, such as energy conservation. The process seems incomplete and poorly targeted, and the categories of the various different subsidies ought to be simplified as much as possible.

Subsidies for the acquisition and rationalization of areas destined for housing development

From 1971 to 1986, allocations from the 1971 law were granted directly to the communes and put into their capital account with the principal aim of creating state ownership of land in the areas destined for subsidized housing schemes.

The grants were distributed by the *Cassa Depositi e Prestiti*, the autonomous (since 1983) public agency for deposits and loans which is controlled by Parliament on the basis of the information provided on a yearly basis by the *Ministero del Tesoro* (the Treasury). The grants were meant for the acquisition of areas included in the plans for social housing (*PEEP*). These areas were subject to compulsory purchase, and until 1980 the owners were compensated at a price slightly above the agricultural land value. During that year the Constitutional Court declared the compensation standards contained in the 1977 law invalid, and from then on compensation payable to owners had to be calculated on the market value of the land. This considerably increased the cost to the community. Assistance was granted via the regions. The state also granted some communes a ten-year interest-free loan for this same purpose: supporting the residential building programme.

Subsidies for residential building

State financial aid Today there is no law that provides for continual aid during the coming years. The ten-year housing programme, approved in 1978, was not refinanced after its termination in 1987. This had allocated 14440 billion lire for the construction of social housing and 1036 billion lire to pay the interest on loans borrowed by companies, co-operatives and public bodies who were involved in the construction and restoration of housing units, which were to be given to mid- to low-income families.

Subsidized building Aid for subsidized building (social housing) are as follows:

- new works dealt with by *IACP*, in areas designated by the commune for social housing (*PEEP*)
- acquisition of houses already habitable and immediately given over to the care of the commune
- the up-keep of the real estate owned by the *IACP* or directly by the municipalities.

The communes have to assign areas to *IACP*, dividing them up between those contained within *PEEP* and the others. It is the job of the commune to assign dwellings through public competition, in which any Italian citizen can participate, for either new or reinstated housing, provided that they are eligible. The primary requirements are:

- not to be the owner of otherwise adequate housing
- to have a family income less than that established by the state and by the regions.

There are other priorities, established by regional laws, dealing with the circumstances in which those seeking help find themselves; for example, enforced eviction, newly regrouped families, pensioners, the disabled, low income, large families, and so forth. The housing must be built to the standards set by the regions. The cost of the construction must not exceed the limits fixed by the state; this limit can be modified only in part by the regions.

Private-sector assisted housing Aid for private sector housing, destined up until 1987 for public bodies, is now reserved for companies and co-operatives for the following:

- property restoration
- construction of housing in areas designated for subsidized housing schemes (*PEEP*), or in other areas belonging to the contractors.

The regions divide up the aid available between provincial or subprovincial areas, giving a quota for restoration work and for public competition for the new buildings, in which companies and co-operatives participate. The aid put towards this competition is divided up in fixed quotas for all interested parties: businesses, full ownership co-operatives and jointly owned co-operatives.

The communes are responsible for assigning areas, according to the same formalities as subsidized housing, with a consequent reduction in assistance. The transfer of the area from the commune to the contractors involves an agreement over the price and amenitites of the area, the amount of the financial contributions paid for urbanization works (roads, sewerage, etc.), the range of selling prices, the siting of dwellings, construction time, and so on. Agreements are also reached among all those concerned with subsidized housing on the size limits of dwellings and construction costs. Contributions towards reduced interest rates vary between contractors who actually build and directly sell the dwellings, and those who are not builders but work in connection with builders to place the properties on the market. They can buy or rent housing for Italian citizens or enable them to become eligible for subsidized housing, provided that their income is within the set limits. With the act of sale, the loan is transferred to the buyer.

"*Buono casa*" (up-front grant for new home-owners) Not all subsidies and grants given directly to citizens for buying, construction and restoration of dwellings are properly covered under this heading. There can be a block-grant to cover partially any such expenses incurred, or a contribution towards paying off the interest on loans for buying; this could be for as long as 15 years. In response to the rise in the housing evictions in the major cities, the state introduced emergency laws in the 1980s. The resultant aid was distributed by the regions or directly to the communes. The eligibility criteria were basically the same as for assisted housing.

The Goria Law The Goria Law started out as a fund which is now all used up, with the *Cassa Depositi e Prestiti* (the public agency for deposits and loans), for employees, with the idea of putting 20% of the salary towards buying a house. See Section 3.2 for a discussion on the eligibility criteria. This law was introduced primarily to benefit families with one income, instead of families with no income at all.

Regional aid This is the same as state aid and the two are usually integrated. Some regional laws determine the subsidies for implementing the integrated programmes proposed by private contractors and approved by the commune. Programme approval is in effect a declaration "for the common good," allowing for the eventual release of planning permits. As well as grants already set aside for ordinary programmes, the communes can receive other contributions towards planning and to cover other necessary expenses for carrying out programmes alongside urban restoration; that is, moving families that are occupying houses about to be restored. Some regions finance the acquisition and restoration of some public housing, having particular regard to its architecture and history (e.g. Piedmont).

Controlled rent

Fair Rent Act A "Fair Rent Act" was introduced in Italy in 1978, to break through the system of frozen rents and contracts that had existed since the war. The Fair Rent Act ensured that, in all cities with more than 5000 inhabitants in 1971, the rental covering domestic use of housing and private property is related to the floor area and type of dwelling. It also ensures that the rent charged is related to a cost of living index (*ISTAT*). The Fair Rent Act is applied to all housing and it pays no regard to the personal circumstances of the tenant. Such strictures mean that an owner cannot summarily evict a tenant just because the date on which an eviction ought to have taken place was put off. The block on evictions has now lasted for almost, and during this period attempts at eviction have generally not been successful.

The law, issued with the aim of putting a price ceiling on the rented property market, created notable distortions:

- Housing contracts changed in favour of alternative usage (e.g. offices, guest rooms, furnished letting), precisely in those areas where the housing need is greatest.
- Owners tended to leave dwellings vacant because the value of the property would fall if it were let.
- Young couples, the old, and families who needed to move often, had to resort to the free market where prices were often unbearable for the weakest sections of the population.

At present, only a small section of the rented housing stock is still protected by the Fair Rent Act as the government approved the *Patti in deroga* in the early 1990s, a deregulation which introduced *de facto* free market conditions into the housing sector.

The social act The self-governing housing associations (*IACP*) were created as public bodies at the beginning of this century in nearly all provinces. They had the task of carrying out building programmes for low-income housing, together with the management and maintenance of the property. The *IACP*s have been replaced by *ATC*s, which are basically the same institutes, but they are fundamentally different in that they have more financial autonomy and thus tend to manage the public housing stock in a more business-like way.

CHAPTER 4
Prices

4.1 Price setting

Available sources of information

Every survey on the Italian real estate market faces many difficulties, owing to the fact that available sources of information are somewhat unreliable. From a statistical viewpoint this is a paradox, since the product to be measured is highly visible and its economic role is so relevant to the country: the construction industry takes up 20% of the total workforce in the industrial sector, and it represents a sizable share of the gross national product. This economically important business is the subject of many surveys undertaken by several different bodies whose aim may be administrative, financial, or statistical. Nevertheless, some of the most important statistics are scarce. This flaw is caused by several factors, the most important of which is the lack of a real estate market in the sense given to this term in other product markets.

It is clear that, in this case, the "product" of the real estate market is divided up on both a typological and geographical basis. Other factors should not be overlooked: for example, global demand and additional supply are localized; there is a lack of information on both demand and supply; many producers only supply the product on a spot basis, and there is a large range of parameters to consider in order to assess the quality of the product.

The degree of heterogeneity and of geographical area is not a "required" feature; rather, it is a variable factor in the housing and planning policies needing implementation. The scarcity of information leads to a narrowing of market options and to increased expenses for potential buyers and sellers. The lack of information is an important cause of high opportunity costs (time lost, etc.) which are added to the price demanded for each property. From the seller's viewpoint, the "low visibility" of the market results in considerable uncertainty over future price trends.

In a real estate transaction one can identify several sources of information: legal, fiscal and financial. However, these sources are of little use in Italy because data is collected for a range of uses that is too wide; data is not comparable because of different data-collection techniques; and data do not refer to market prices but to other values, for example the "price declared" by buyers and sellers in the presence of a notary public, or the property value as estimated for granting a mortgage. This is why in Italy the most reliable information source on actual market prices is an expert opinion on "typical prices" for various categories of property.

As a consequence, the difference among prices given by different sources has to be interpreted as the sum of endogenous and exogenous factors. "Endogenous factors" here refer to the great heterogeneity of properties, whereas exogenous factors refer to the statistical nature of the sample utilized as the source of information. For example, according to *Monitoring service of property markets* (*Osservatorio sul mercato immobiliare, Nomisma*) the lowest housing prices in central Milan increased by 24% from 1989 to 1990, while the top prices increased by 40%. But according to other sources, the increases were respectively 56% and 60% between 1989 and 1990. The difference is caused by statistical factors, rather than real market changes. This research has been conducted in other cities with similar results. In conclusion, it is necessary to take a cautious view of estimates of property prices in Italy.

The urban housing market

When analyzing house prices over a period of time, it is advisable to separate values from the dynamics of inflation. The price trends between the early 1960s and the mid-1980s were marked by two periods: first, the oil crisis in 1973 and, secondly, the crisis of 1978. These turning points, the second of which was more important, were marked by a sharp increase in actual prices, then by a subsequent decrease, which only partially compensated for the previous rise. On both occasions the number of house sales rose enormously, fuelled by a widespread climate of uncertainty fostered by high inflation rates and bad economic judgement. The attention of householders was drawn to home-ownership, which was seen as the most secure form of investment.

Through the 1980s, house prices performed a U-turn. The decade started with a strong increase in values, but the trend had already effectively reversed in 1982 when a long phase of "steady prices" began. That meant declining prices in real terms, as stable prices were continuously eroded by high inflation rates. The prices did not start to rise again until 1987, with a true price-boom in 1988–9. But in this case new house sales were motivated by a new rationale: the demand was particularly intense for historic buildings located in prime urban areas (usually central areas). The "willingness to buy" arose from the concentration on housing quality, which is a key factor in a market where well over 70% of the households are already homeowners: it is estimated that in 1980, out of every ten housing transactions, seven sales concerned first-time homeowners. In 1990 the proportion of first-time homeowners was only two out of ten, and it is still declining.

Since the early 1990s, housing prices have been declining, and this downturn marked a new phase in the property markets in Italy. A particular feature of the property market in the main Italian cities is that values in different neighbourhoods show different patterns. These differences are obvious when housing prices located in central areas are compared with those in peripheral areas. Figure 4.1 is somewhat exceptional, because the data spans a long period of time (1965 to 1995) and comes from the same source, the Real Estate Consultant (*Consulente Immobiliare*). This journal, published by and for property professionals, is one of the most reliable sources of information on property prices in the country. The

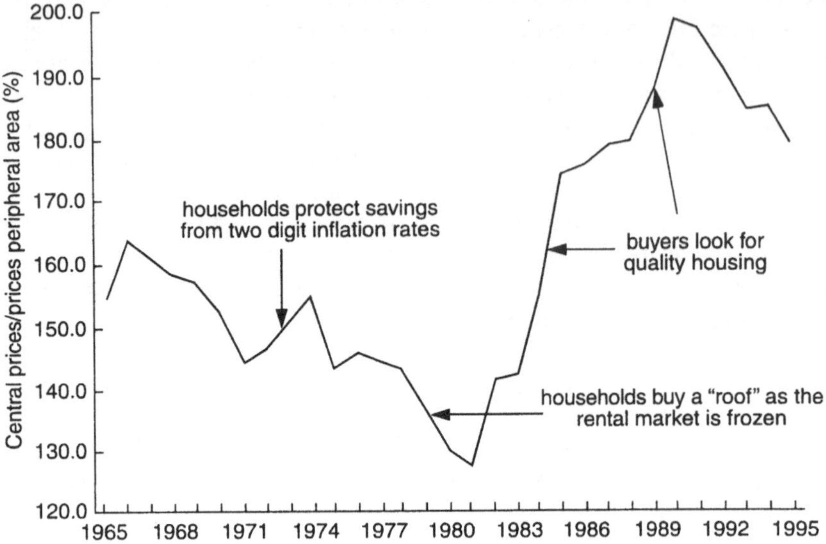

Figure 4.1 Prices ratio centre–peripheral areas new and non-occupied housing units.

journal produces a wide range of data on real estate properties twice a year. In the first years of publication, the data covered about 40 cities, whereas in the past few years data collection has been extended to include the prices observed in 104 cities.

During the period 1965–75 there was a slow but constant decline of prices in central areas (the price ratio between central and peripheral areas fell from 2 to 1.5), because of a substantial increase of prices in the peripheral areas combined with a widespread decay of the inner city. The decline of the centre–periphery ratio can be explained for the 1978–81 period as the market outcome of the Fair Rent Act, which was approved in 1978. As a result of the law, the rental market froze and the households who were looking for accommodation had almost no other choice than to buy a house, almost any house (in the sense of a roof), primarily in the outskirts of the city. This pushed prices up, in spite of the poor locations and low-quality housing. But, starting in 1981, the relationship reversed and central prices became on average 2.3 times greater than those of comparable properties in peripheral areas. Demand for housing and offices in central locations grew compared to that for peripheral areas, which because of traffic congestion were difficult to reach from the city centre. These features can be found in other countries, and in Italy are the result of structural changes that affected the economy. Without adequate provision for efficient transport and telecommunications networks, it seems unlikely that there will be any change in the continuing process of centralization. Since it is difficult to reach city centres from far away, a site in the heart of the city becomes a necessity. The price rise of central areas can also be explained by the fact that, in the 1980s, it became easier to

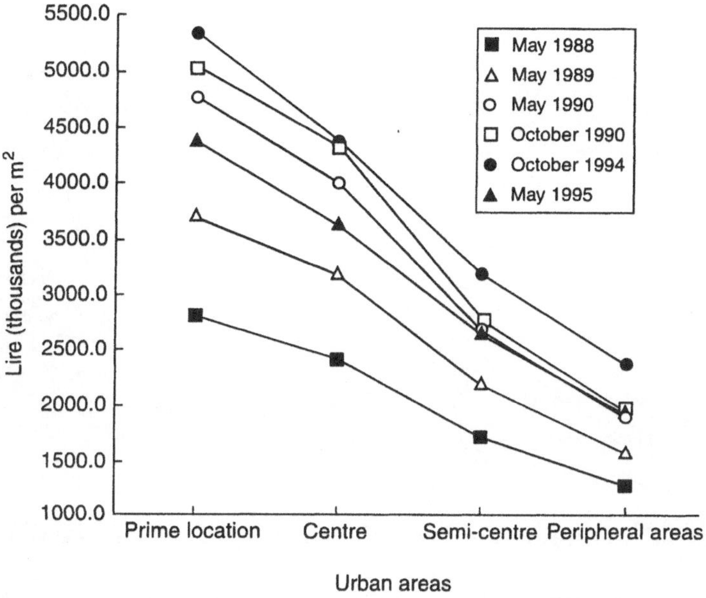

Figure 4.2 Average prices in the 13 largest Italian cities for new (or completely restructured) and non-occupied housing units, by urban location.

launch extraordinary refurbishment projects in the city centres. The law of 1978 authorized for the first time limited intervention in city centres through an authorization (*autorizzazione*), which was granted with the so-called "silent-approval" method. A no answer within 90 days from the request of the building permit meant "approved" instead of "rejected". Between 1990 and 1995 there was a sharp decline (from 2.0 to 1.8) in the centre–periphery prices ratio, which can be interpreted as a market response to the frantic price increases of the late 1980s, which had taken place primarily in the city centres (see Figure 4.2).

This relationship can be interpreted as a rough guide to urban rents, given the lack of data on urban land prices. It is worth noting that in Italy no reliable data exist on the prices of urban land suitable for development. The only available information is the estimated percentage share of land value over the total property value (see Figs 5.8–5.11) and again the data has been obtained from the *Consulente Immobiliare*.

In the case of housing for primary use, the available data is impressive for the length of the time period and for the geographical area covered: 32 years (1963–95) for all the 13 largest Italian cities (Bari, Bologna, Cagliari, Catania, Florence, Genoa, Milan, Naples, Padua, Palermo, Rome, Turin, Venice), with prices expressed at constant values (1980 lire) in millions per m^2 of floorspace, and subdivided in three urban zones (prime centre, average centre or mid-town, and peripheral areas).

Figure 4.3 has been drafted on the basis of the data. It shows a single line

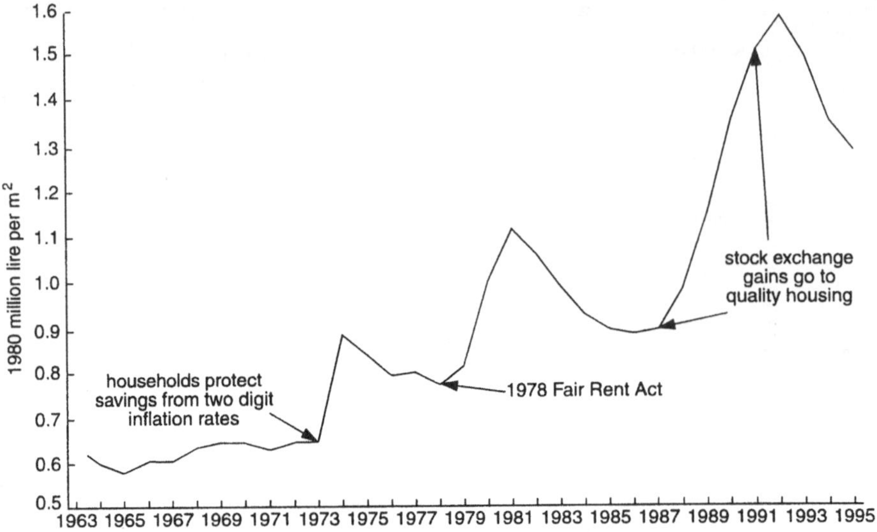

Figure 4.3 Average housing prices for new and non-occupied housing units in large Italian cities. Constant values (1980 lire).

which represents the average housing price for all 13 largest Italian cities for the period 1963–95. It shows an exceptionally regular trend, with three peaks corresponding to the years 1974, 1981 and 1992 when the price index at constant 1980 lire (base 100 in 1963) reached, respectively, 142, 178 and 250.

The first peak of prices (1974) was produced by the effects of the first oil shock, which led Italy into a two-digit inflation nightmare from which the country recovered only in 1984–5 when the inflation rate eased from 10.6% to 8.6%. In order to protect their savings, many households invested in housing, marking a first massive wave from rental housing to homeownership.

The second impressive "migration" to homeownership occurred after the approval of the *Equo canone* (Fair Rent Act) by the first and only Italian government led by the Christian Democratic party, supported by a friendly opposition of the Communist Party, in 1978. The law was successful in providing clear rules and protection for the households already in the rental sector. But it discouraged new investments in the private rental sector because it set a cap on the rents that owners could charge without breaking the law, and because it made it very cumbersome for the owner to terminate a rent contract and reclaim the property. Housing for rent soon disappeared from the market and all those in need of a house (e.g. young couples) had to choose between paying two rents, one according to the amount established by the parameters indicated by the law and one "under the table", or becoming homeowners. Hence, there was a surge of demand for housing of almost any quality and in any location.

The outcome of the favourable economic conditions in the late 1980s was strong new demand for housing. But most of the new buyers were already homeowners and belonged to the medium and upper-medium income levels. Consequently, the demand for housing was completely different from the one at the end of the 1970s caused by the disappearance from the market of housing for rent. It was a demand for houses of higher quality (including better locations) and more space than the houses already owned. The price index climbed from 142 in 1986 to 250 in 1992.

It can be observed from Figure 4.3 that, following each peak, there was a decline of the average price as rapid as the previous rise. Most Italian promoters and even real estate experts failed to recognize in the early 1990s that the housing market was on the edge of a recurrent downturn, in spite of some serious warning signs during the late 1980s, notably the increase of the average time required to close the sale of housing properties. They simply preferred to repeat the old (false) story that housing prices would move only in one direction (upwards), until the Italian cities were flooded with "for sale" (*vendesi*) signs and frustrated owners realized with dismay that their properties would not sell, even after the first price reductions.

The available data shows that, on average, the decline in house prices was steady until 1995 when the index reached 205. As of 1996 it does not appear that the market has bottomed out. According to the observations made for the previous three price peaks, it would have been a surprise if the price downturn had recovered immediately. The analysis of the past seems to provide evidence that the average price trend will not reverse until 5–7 years after the 1992 peak; thus, average housing prices should increase again, starting in 1997–9. By the same token, a new price peak may be expected in a period of between 6 and 11 years from the previous peak, thus in the 1998–2003 period.

The urban market in business properties

In Italy the 1980s saw rapid growth in the service sector of the economy. Many industrial firms moved from the few central locations to new areas, whereas the service sector rapidly increased its demand for properties. This structural change, combined with a morphological change in the city, led to strong demand for office space. This trend might receive a further impulse from the development of the Single European Market, which entails, among other things, a growing demand for office space in the main economic and financial centres of the European Union.

In the late 1980s, after the previous surge of demand for office space, many private speculators entered the office market on the grounds that this market would guarantee good economic returns in terms of both income and capital gain. This resulted in an over-supply of office space similar to, but on a smaller scale than, the over-supply experienced in other European countries. Office prices, which had been constantly rising with housing prices, came to a standstill because of this increase in supply, which (naturally) acted as a regulator of prices and subsequently of office rents.

In 1990 the urban property market for office space stagnated. The number of sales and rental contracts remained stable in all Italian cities, with the possible exception of peripheral areas, where there was a potential demand for large modern office blocks. According to *Monitoring service of property markets* (*Osservatorio sul mercato immobiliare, Nomisma*), Milan ranked first with an average price of 10 million lire per m^2 for office space, followed by Rome (8 million lire), Venice (5 million lire), and Florence and Bologna (4.5 million lire). At the bottom of the scale one can find the cities of Palermo and Cagliari (3 million lire) and Catania (2.5 million lire). It is worth noting that in each city these prices are paralleled by average house prices.

As far as the rental market is concerned, there has been a "time gap" in the slowing down of rent levels compared to the decrease in office sales. Yearly gross revenues for office investments decreased from 8% in 1988 to 6–7% in 1990, whereas the price range across urban areas narrowed and came closer to central values.

The urban retail property market has followed a trend similar to the housing market. Its peculiarity consists of a greater variety of properties in terms of location and floor area. Supply was lower than demand and this created tensions in the market, reflected in high prices. Table 4.1 shows the price trend of retail properties and offices in the 13 largest Italian cities between 1988 and 1995. The average price of retail properties increased swiftly from 2.4 million lire per m^2 of floorspace in 1988 to 3.9 million lire in 1992; a rise of 67% in four years. Since then, the average prices fell constantly to 3.3 million lire per m^2, while the index (1988 = 100) fell to 141 from the peak of 167 in 1992. The sharpest reductions occurred in Milan and Rome where price levels had attained unprecedented heights in 1992. The trend for office prices followed a similar pattern. The index of average prices with the same base (1988 = 100) reached the highest values in 1992 (186) and 1993 (189), and has been falling ever since. In 1994 the analysis of actual office prices in the 13 largest Italian cities shows values much closer to the average than in the late 1980s. The sharpest variations took place in Rome and Milan, where prices had exceeded 4 million per m^2 of floorspace in several years. Milan and Rome, once again, were the most speculative office markets of the country. By contrast, in other major office markets, such as the cities of Turin, Bologna and Naples, prices rose and fell less sharply. This is a characteristic that may not make the financial headlines, but should be seen as an interesting feature for those investors, such as pension fund managers, who seek a stable return rather than the quick speculative gain (which may result in a speculative loss).

In statistical terms, there is a definite link between top and bottom prices of housing, office and retail properties. Minimum and maximum prices change accordingly. When minimum prices increase, maximum prices also increase, and so on. The covariance index of minimum and maximum prices is so high that it approaches a perfect linear relationship. This provides useful information on the types of urban properties. In the case of housing, the price of prime properties shows a constant relationship with the price of lower-quality housing, in terms of both location and physical standards.

Table 4.1 Weighted average selling price (millions of lire per square metre) of office and retail properties in the 13 largest cities.

	May 1988	May 1989	May 1990	May 1991	May 1992	May 1993	May 1994	May 1995
Retail properties in commercial areas								
Bari	2.381	2.793	2.998	3.294	3.479	3.337	3.042	2.887
Bologna	1.863	2.680	3.163	3.707	3.801	3.864	3.602	3.428
Cagliari	1.938	2.374	2.644	2.856	3.281	3.269	2.868	2.726
Catania	1.710	1.882	2.395	2.482	2.565	2.600	2.527	2.554
Florence	2.764	2.950	3.742	4.083	3.883	3.948	3.438	3.521
Genoa	1.901	2.390	2.836	2.945	3.489	3.647	3.268	3.264
Milan	2.579	4.019	5.748	6.408	6.196	5.496	4.695	4.362
Naples	3.815	3.815	3.865	4.201	3.927	4.225	4.573	3.942
Padova	2.077	2.457	3.076	3.236	3.490	3.512	3.286	3.282
Palermo	1.993	2.313	2.659	3.108	3.469	3.445	2.740	2.538
Rome	3.346	4.876	6.233	6.396	7.136	6.298	5.040	4.519
Turin	1.756	1.997	2.613	2.843	3.160	3.100	2.655	2.492
Venice on lagoon	3.455	4.193	4.642	4.758	4.812	4.807	4.445	4.778
Venice on ground	1.492	1.660	2.258	2.498	2.476	2.805	2.486	2.492
Average	2.362	2.886	3.491	3.773	3.940	3.882	3.476	3.342
Index	100.0	122.2	147.8	159.7	166.8	164.4	147.2	141.5
Office properties								
Bari	1.843	2.120	2.185	2.599	2.712	2.567	2.348	2.118
Bologna	1.565	2.305	2.628	3.128	3.239	3.353	3.284	3.327
Cagliari	1.218	1.552	1.623	1.848	2.195	2.340	2.121	2.064
Catania	1.007	1.048	1.420	1.583	1.768	1.939	1.733	1.827
Florence	1.827	1.955	2.899	3.150	3.348	3.424	3.251	3.117
Genoa	1.353	1.747	2.138	2.264	2.630	2.571	2.325	2.481
Milan	1.905	2.818	3.765	4.467	4.229	4.057	3.860	3.470
Naples	1.603	2.016	2.349	2.520	2.445	2.655	2.637	2.651
Padova	1.326	1.827	2.415	2.478	2.604	2.541	2.374	2.457
Palermo	1.110	1.435	1.606	2.087	2.286	2.216	1.942	1.823
Rome	2.337	3.150	3.778	4.100	4.078	4.217	3.560	3.232
Turin	1.169	1.640	2.095	2.550	2.795	2.809	2.593	2.463
Venice on lagoon	2.152	2.858	2.979	3.106	3.451	3.514	3.380	3.613
Venice on ground	1.041	1.413	1.673	1.927	2.037	2.386	2.180	2.233
Average	1.533	1.992	2.397	2.701	2.844	2.899	2.685	2.634
Index	100.0	130.0	156.4	176.2	185.6	189.2	175.2	171.9

Source: Nomisma, Osservatorio sul mercato immobiliare **2**, 1995.

Interesting patterns can be seen from an analysis of the linear correlation between top and bottom prices. Examining the performance of top prices in relation to bottom prices, one can see that, when minimum prices increase or decrease by 1 lira, maximum prices tend to increase or decrease, on average, by more than 1 lira. This pattern is to be found in all housing submarkets: new housing units, used housing units and housing for renovation. For office and retail properties, top price variability is even greater (when bottom prices increase or decrease by 1 lira) than for housing. This means that top prices are more sensitive to any variations in bottom prices.

There are several reasons for this. Rents are still partially controlled in the housing sector, while rents are basically determined by the market for retail and office properties. Investors have more options in the office market than in any other property market. Finally, office and retail properties can be considered as "business premises" and as such are eligible for tax relief.

4.2 The actors and their behaviour

An overall picture of real estate development in Italy can be gained from an analysis of investment in the various areas of the construction industry. In 1986, investment in housing was 47% of the total (24% new construction, 23% refurbishment), 25% was directed to non-residential buildings and 28% to public works (Table 4.2).

Table 4.2 Fixed investments in the construction sector.

	Residential constructions			Non-residential constructions and public works			
Years	New	Restoration	Total	Constructions	Public works	Total	Total fixed investments
1980	27.3	20.2	47.5	27.2	25.3	52.5	100.0
1981	27.5	20.1	47.6	27.2	25.2	52.4	100.0
1982	26.6	20.4	47.0	26.2	26.8	53.0	100.0
1983	26.0	20.7	46.7	25.6	27.7	53.3	100.0
1984	25.5	21.1	46.6	25.5	27.9	53.4	100.0
1985	25.0	22.0	47.0	25.3	27.7	53.0	100.0
1986	26.0	22.8	48.8	24.3	27.0	51.3	100.0
1987	25.1	23.1	48.1	24.3	27.6	51.9	100.0
1988	24.7	22.4	47.0	26.2	26.8	53.0	100.0
1989	24.7	21.5	46.1	27.1	26.8	53.9	100.0
1990	25.0	21.6	46.6	28.2	25.2	53.4	100.0
1991	25.0	22.0	46.9	28.5	24.6	53.1	100.0
1992	25.7	22.5	48.2	27.2	24.6	51.3	100.0
1993	27.2	23.7	50.9	26.6	22.5	49.1	100.0

Source: CRESME.

Within the public sector, investments can be subdivided according to the promoting authority, although it has to be kept in mind that in Italy the responsibility for the government of cities falls with local authorities: there are 20 regions, 103 provinces, 8103 communes (Table 4.3). Spending on public investment in 1985 shows that provinces and communes were the main spenders (45%), followed by local public companies (12.5%), central state and social security (10.4% each). In the period 1985-94, the spending share of provinces and communes decreased to 41.7% in 1990 and 42.2% in 1994. Conversely, the regions increased their spending from 8.8% in 1985 to 12% in 1993 , while their share returned to 10.6% in 1994 (Table 4.4). Spending figures for public investment in 1994 show that

Table 4.3 Communes, gross area of forestry and agricultural land as of 1993 (areas in km^2).

Regions	Number of communes	Number of provinces	Gross territory Total	%	Gross forestry and agricultural land % on gross territory
Piedmont	1209	8	25399	8.4	86.2
Aosta Valley	74	1	3262	1.1	65.7
Lombardy	1546	11	23857	7.9	79.2
Trentino A. Adige	339	2	13618	4.5	84.3
Veneto	582	7	18364	6.1	80.5
Friuli–V.Giulia	219	4	7845	2.6	79.4
Liguria	235	4	5418	1.8	87.9
Emilia–Romagna	341	9	22123	7.3	86.6
Tuscany	287	10	22992	7.6	89.4
Umbria	92	2	8456	2.8	92.4
Marche	246	4	9693	3.2	91.2
Lazio	377	5	17203	5.7	86.5
Abruzzo	305	4	10794	3.6	91.3
Molise	136	2	4438	1.5	92.3
Campania	551	5	13595	4.5	87.5
Puglia	258	5	19348	6.4	91.0
Basilicata	131	2	9992	3.3	92.6
Calabria	409	5	15080	5.0	89.3
Sicily	390	9	25709	8.5	90.0
Sardinia	376	4	24090	8.0	92.4
Total in Italy	8103	103	301277	100.0	87.3

the provinces and municipalities were the main spenders (42.2%), followed by central state companies (*ANAS*, etc. 14.6%) (Table 4.4).

In the new-housing sector, planning permission was granted mostly to private individuals (43%), followed by companies (30%), cooperatives (17%) and public institutions (10%) (Table 4.5). It is worth emphasizing the power of building firms: traditionally they not only build but they are also developers, in various degrees. In small and medium-size towns, building firms act in collaboration with the local professionals (mainly architects and *geometri*, (quantity surveyors)) who possess information on land development opportunities. However, in the main cities developments are increasingly negotiated with the local authorities.

The role of housing cooperatives, which act as a pool supplying individual demand, is also important. Each cooperative consists of at least nine members who have the common goal of constructing houses for their own private use. There are two kinds of housing cooperatives: those with undivided assets and those with subdivided assets. In the former the finished units are owned by the cooperative, which assigns them to its members; in the latter the housing units are owned on an individual basis by the members of the cooperative.

As far as housing refurbishment is concerned, there is no precise data. However, it is estimated that the vast majority of it (more than 70%) is done by private individuals. On the other hand, larger-scale work is undertaken by firms, real estate companies, insurance companies, banks and professional developers.

Table 4.4 Spending on public investment (billions of lire).

	1985	(%)	1986	(%)	1987	(%)	1988	(%)	1989	(%)	1990	(%)	1991	(%)	1992	(%)	1993	(%)	1994	(%)
State	2802	10.4	3540	2.2	3921	12.3	4252	12.2	4639	11.9	4983	11.6	5705	12.2	5261	11.6	5206	12.6	5278	13.8
Central companies	2685	9.9	3074	10.6	2945	9.2	3380	9.7	3892	10.0	5904	13.7	7152	15.4	7152	15.7	6422	15.6	5563	14.6
Regions	237	8.8	2805	9.7	3487	10.9	3939	11.3	4340	11.2	4921	11.4	5308	11.4	5271	11.6	4940	12.0	4034	10.6
Provinces and communes	12146	45.0	13094	45.2	14402	45.1	15676	45.1	17548	45.1	17987	41.7	17928	38.5	17843	39.3	16527	40.1	16068	42.2
Hospitals and *USL**	821	3.0	1078	3.7	1248	3.9	1413	4.1	1699	4.4	2373	5.5	2403	5.2	2045	4.5	1813	4.4	1567	4.1
Local public companies	3388	12.5	3686	12.7	3686	11.6	4162	12.0	4346	11.2	4251	9.9	4713	10.1	4315	9.5	4109	10.0	4287	11.2
Social security bodies	2804	10.4	1666	5.8	2221	7.0	1962	5.6	2439	6.3	2722	6.3	3378	7.3	3554	7.8	2152	5.2	1314	3.4
Total	27020	100.0	28943	100.0	31910	100.0	34784	100.0	38903	100.0	43141	100.0	46587	100.0	45441	100.0	41169	100.0	38111	100.0

Source: Ministero del Bilancio. * Local health units.

Table 4.5 Division of total housing between the holders of building concessions.

	State and local government bodies	Other public bodies	Firms	Housing co-ops	Private and others	Total
Triannual 1980–83	2.2	7.0	30.7	19.8	40.3	100.0
Biannual, 1984–5	1.7	8.6	29.9	16.8	43.0	100.0
1986	1.0	7.9	30.4	16.3	44.4	100.0
1987	1.5	5.4	34.4	15.2	43.5	100.0
1988	0.8	3.8	35.0	15.8	44.6	100.0
1989	0.6	3.3	37.2	11.1	47.8	100.0
1990	0.5	2.2	38.3	12.6	46.4	100.0
1991	0.4	2.1	37.1	12.1	48.3	100.0
1992	0.5	2.1	37.6	11.2	48.6	100.0
1993	0.4	2.9	38.3	10.7	47.7	100.0

Source: ISTAT.

In the non-housing sector the most significant components are service- and shopping centres and industrial buildings. The former and quickly expanding sector is the target of the main developers: financial companies, building firms and insurance companies. They usually "assemble" the operation and negotiate with the local authority. In practice, the operation is mainly handled by smaller developers and *geometri* (quantity surveyors). In the medium-size and small towns, *geometri* often play the role of development experts, in both housing and industrial development, because of their knowledge of the local real estate markets.

In the building sector in Italy three kinds of professionals are officially recognized: architects, engineers and *geometri*, who are organized in professional institutions. It should be noted that the "*geometri*" are a profession very typical of Italy. They make real estate evaluations and are also entitled to "sign" projects and to request, on behalf of the property owner, building permits for works below a certain financial threshold (above which the project has to be signed by an architect or by an engineer).

The professional institutions (*Ordini*) are associations instituted by a royal decree in 1925. Membership is obligatory for anyone involved in an activity that requires specific abilities and carries specific responsibilities. Professional institutions have a quite limited range of action, usually corresponding to the provincial territory. There is also a central body, the National Council, based in Rome; it represents the profession as a whole and has advisory and decision-making functions.

As of 1995, the above-mentioned professional institutions have declared the following membership figures: architects, 70000 members; engineers, 111000 members; *geometri*, 82000 members. Finally, professional intermediaries, that is firms that deal with real estate transactions, exist but are less developed in Italy than in other European countries. There are no detailed statistics, but it is estimated that in the smaller towns around 70% of transactions are done without having recourse to "intermediation services", and in the main cities one can find a larger percentage of sales carried out with the help of a real estate agent. Another specific feature of the Italian market is the presence of many operators. It is estimated that there were around 70–80000 estate agents in 1995, a number that increases or decreases with the real estate market cycles.

Before 1989 the business activity of real estate intermediaries was not regulated by any specific norm. Individuals interested in becoming real estate agents could do so by simply registering with the local chamber of commerce and the local finance office of *IVA* (the value added tax). Fraud was not uncommon. The newspapers of large cities were sometimes flooded with stories of dishonest agents selling the same house to several citizens before vanishing with the cash. Apart from unlawful cases, the presence of many occasional operators, with their sometimes poor professional standards, has helped to create a negative image of the profession.

In 1989 a law was approved to regulate the intermediation activity in the real estate sector. For the first time the law defined the content of the activity of *agente immobiliare*, real estate agent, and its minimum professional qualifications. A roster of official agents was set up in each chamber of commerce. Enrolment to this roster is granted to new agents only after a training course of about 100 hours, held by the professional associations of real estate agents. The law required the submission with the local chamber of commerce of standard forms of contract, which specified the maximum percentage fees to charge to clients.

According to the law, those operating without a licence will loose their right to any commission. Agents charge between 4–8% of the estate value as intermediation fees; costs usually being borne equally between buyer and seller.

Nowadays, there are some well established firms (for example, *Gabetti*, *IPI*, *Tecnocasa*) who offer high quality services. The birth and the development of professional associations, such as *FIMAI* and *FIAIP*, have also gone a long way towards setting up formal qualifications, and thereby to offer some kind of guarantee to clients. In recent years larger real estate companies have widened and diversified their activities, to the point of actually promoting complex development ventures.

4.3 Implementing the planning process

The municipality is the only authority that can grant planning consent, and decisions must be consistent with the overall town plan.

Let us now analyze some examples:

- A building requires consent according to the city master plan (*Piano regolatore generale*, *PRG*) or building plan (*Piano di fabbricazione*, *PF*). In this case, consent can be obtained as long as the building conforms to the zoning and the development plan for the site.
- The site may be listed as one requiring an action plan, in which case there are two possibilities:
  - The action plan is drafted by a private entrepreneur or developer and covers a wider area than the specific site for which permission is being sought. If the action plan has already been drafted, the procedure is as outlined above; if not, an action plan is required. The various land and property owners will have to submit a proposal, as they are the only ones entitled to do so.

In the event of there being more than one owner, they will have to agree to a joint action plan. This action plan must then be approved by the municipality, and according to local regulations it may have to be approved by the region too. Once approved, each owner (if there is more than one) can apply for building permission and the procedure follows that described above).

  - There is also the case of a "public action plan", that is to say a plan drafted directly by the municipality, which in some areas may also have to be approved by the region.

There are several kinds of action plans. The plan for public housing in residential sites is known as *Piano per l'edilizia economica e popolare* (*PEEP*) or *Piano di zona* (*PZ*) if the local authority has previously bought the site. The sites are initially divided up into plots and then "urbanization" work is carried out, i.e. the area is supplied with the necessary infrastructure: roads, connections and the services that the development may require. The sites are subsequently allocated. A portion of the area may be sold leasehold, with a lease not exceeding 99 years, to be renewed for another 99 on request. The remaining property can be sold freehold to agents operating in the field of subsidized housing. In both cases there are charges and transfer taxes. Municipal authorities can set prices for both leases and freeholds for the original site.

Development and management consortia or corporations are often created and act according to the agreements signed with the local authority. The latter may choose to delegate all the development to these consortia. In many cases the creation of these *PEEP*s is the result of the indirect regional incentives and they fall within the scope of their authority.

In any event, the municipality is funded by the region to set up the residential estate (i.e. purchase the land and start the site work). Regions operate through ad hoc public financial institutions, which in turn may choose to act by creating jointly owned companies with local authorities and with other institutions.

On the whole, the availability of residential areas allocated to *PEEP*s in northern Italy is below the actual demand, especially in large metropolitan areas, whereas the southern parts of the country still offer interesting opportunities because of the special government funding of the development and aid agencies operating in the South.

CHAPTER FIVE
Outcomes of land and property markets

5.1 Changes in the structure of ownership

At a national level the privately owned housing in Italy accounted for 75.5% of the total stock in 1993 (*ISTAT*), and it further increased by some 0.5% by 1996. This is quite a rapid rise from the levels of previous decades: homeownership was still 53.4% of the total in 1961 and 55.8% in 1971 (see Table 1.17). The trend outperformed the experience of other EU countries, which on average saw the quota for privately owned houses pass from 40% in the 1960s to its current level of 59%. The national value masks great differences between the North and the South of Italy and, above all, between the large cities, where the homeownership rate is lower, and small centres, where it is higher. Most of the rented houses are located in the 13 metropolitan areas of Italy (from Turin to Palermo) that are usually referred to as the parts of the country under more or less continuous "housing pressure". Nevertheless, the North has a higher rate of rented housing than the South, across all city sizes.

The tendency of privately owned housing to grow continues for various reasons: the return of real investment in housing after a period of preference for the commercial real estate market; the growth of income and of savings per capita; the great reduction of new public housing initiatives, after the completion of the *Piano decennale per la casa*, (the Ten-Year Housing Programme); the limited availability of private lodgings for rent; and the inflated real market prices of lodgings created by speculative activities. In 1981, in some large metropolitan areas, there was still a balance between rented houses and privately owned houses. For example, in Turin and Milan four out of ten houses were rented, a proportion much higher than the national average. In the 1980s there was a decline in population in both cities. Generally, the people simply moved just over the commune borders into the surrounding localities of the respective metropolitan area. It appears that those who left were mainly tenants, so the incidence of homeownership in the two cities increased and it approached the national average by the mid-1990s.

5.2 Demand and supply of property

In the course of the 1980s and 1990s there have been significant demographic changes in Italy. These have had an impact on the demand for housing (in terms of both quantity and quality) and have contributed to the relationship between cities of different sizes and their hinterlands. The population in the central metropolitan areas is falling, whereas it is growing in the suburbs (only in Rome and Naples is there a positive balance in the overall metropolitan areas). The phenomenon reflects not merely a spreading out of the population, but a redistribution of social groups within the locality. The processes of suburbanization interest the middle classes, but at the same time there is a process of selective reurbanization by certain social groups towards the central and semi-central areas of some large metropolitan zones.

The thesis today is that the most dramatic conditions of quantitative need for housing have now passed and there is a transition towards a phase in which qualitative aspects of housing and locational preferences tend to prevail. Within this phase the important issues will be the rate of aging of the population, the composition of households, changes in life-styles, and immigration.

A revealing and general growth in the number of households is taking place. It is caused by the decline in the number of component members of households, the increasing number of middle-aged people, the affirmation of the "single" family and the phenomenon of immigration. These factors (summarized in Table 5.1) still determine the quantitative demand for housing.

Table 5.1 Characteristics of housing demand, 1960–90.

Housing quality	Housing demand 1960–80	1980–90	Expected trend	Actual market supply in the 1990s
Top	Weak	Strong	Fairly strong	New housing and renovations in prime central locations and selected external areas
High	Sustained	Very strong	Fairly sustained	Renovations and new housing in good locations
Average	Sustained	Weak	Fairly weak	New housing on the outskirts
Fairly low	Sustained	Weak	Very small	Weak maintenance, old buildings in low quality outskirts
Sub-standard	Strong	Sustained	Strong	Vacant buildings and areas of new immigrants

The demand for non-residential property has followed the economic cycle: companies initially engaged in phases of technological restructuring of productive processes and subsequently in the processes of enlargement and modernization of their plants. The service sector presents a consistent and growing alternative for the re-use of broken-down industrial areas.

Construction financed with public funds has declined. This has meant reduced business opportunities for enterprises that specialize in pure construction, whereas the firms that have operated as real estate promoters have increased their activities. In general terms, in the housing sector, there has been a constant decline in the percentage weight of "new construction" in total added value and a corresponding surge in "maintenance and renovation". This phenomenon has become evident since 1987–8. In the past decade, in most large cities, the number of true new housing projects has been negligible in comparison to the large numbers of converted units and completely renovated buildings. It is likely that renovation activities have been sustained by local policies keen to counteract the flight of residents from large central communes to suburban locations. It is also likely that the financial resources invested in these properties came from both low-income households and the upper middle class. In the case of the former it consists of those who had dwelled longer than expected in public housing schemes, that is after the improvement of their income and social conditions would have enforced their exclusion from the list of those eligible for social housing. By remaining in their social housing they were able to accumulate savings which were then invested in private real estate. Within this group there are also households who occupied a "frozen rent" house from the post-war period until 1978 and a "fair rent" house after the enactment of the national rent law in that year. There is an irony about the two social measures (social housing and rent freezing or rent control), which in the end obtained exactly the opposite of the stated goals: a sizeable share of the assisted households changed their social condition from house renters to house owners, not because of a free choice but simply because they took advantage of their privileged housing position. This passage (from renting to ownership) frequently took place in the outskirts of large cities and in the less valued parts of the city centres. The houses purchased were in poor condition. Investing in maintenance was also a means, for the new homeowners, to mark their attained social difference from those left behind in the rental market.

In the case of the latter it includes the middle and upper middle-class households who, during the second half of the 1980s, for the first time ever in Italy made financial fortunes on the Italian stock exchange. Homeownership was already secured by these households. The wealth, accumulated in a short time and with luck in the years 1985–7, was invested in higher-quality housing in prime locations. In Italy the best locations are usually the city centres where, obviously, there are very few opportunities to buy a brand new house, so, the new wealth ended up financing the activities of the real estate promoters who specialized in buying and renovating old buildings in central areas.

In parallel, starting in 1985, there has been an increase in the production of non-residential buildings, especially those linked to the service sector. Overall, this sector seems likely to carry on growing, but it comprises various different subsectors which show contrasting growth trends that would require an ad hoc analysis.

The retail market is strong in the centres of the large cities, where demand clearly continues to exceed supply, whereas the opposite is beginning to apply in the areas outside the centre. This is the outcome of the spread of extensive shopping centres during the 1980s and 1990s. The hotel sector is currently expanding, with a strengthening of international chains. After an expansion in the second half of the 1980s, the office sector is going through a crisis. The top values reached in the past decade have generally declined in the course of the 1990s. Only in the true prime locations of the centres of major metropolitan Italian cities (notably Rome and Milan) can values much greater than the average be found.

5.3 Sale and rental price levels

In Italy there is no single real estate market, but rather many submarkets which do not communicate between themselves. They are segmented by different geographical areas of the country, different cities, and by different categories of buyers. The market prices are determined by positional values which take into account the characteristics of each location (primarily accessibility) and the specific qualitative features of each real estate property. In the most speculative markets, prices also reflect the expectations of a general improvement of property values resulting from the outcome of massive public spending programmes in urban infrastructures. This is the case, for example, in the rise of real estate values that occurred in certain neighbourhoods of Milan and of San Donato Milanese as the result of the construction of Line 3 of the city underground. In a general perspective, this is likely to be the case for Rome, where for the year 2000 much public investment is planned to ameliorate urban transport and local public services. This is in connection with *Giubileo 2000*, the event that will attract Christians the world over and which is likely to bring to the "Eternal City" some 30–40 million visitors. Likewise, property values in Turin will certainly benefit from the planned high-speed railway link with Lyons and Paris, on the western border, and with Milan, Venice and Rome to the east and south. Once the general public realizes that the project will greatly increase the accessibility of Turin with regard to the rest of Italy and the centre of Europe, the property market will follow suit.

A comparison of actual values of the sale prices of housing and shops over the course of fifteen years can be developed with reference to the five leading cities in the large metropolitan areas of Italy (Turin, Milan, Genoa, Rome and Naples) and also to five other large Italian cities (Florence, Bologna, Bari, Venice and Palermo). The data is significant, above all when it is used for comparison between cities, because some caution is suggested with regard to the absolute real estate values which are probably biased. It should be noted that most of the following graphs have been based on constant values (1980 lire) to highlight pure market trends without the distorting bias of monetary inflation. Obviously, actual values are much higher, as is shown in the tables with figures at current values.

As indicated by Figure 5.1, average new housing prices in the central areas of Rome and Milan followed a similar trend. After a downturn between 1981 and 1985, prices rose sharply in the second part of the 1980s, to reach record high levels in 1991 and 1992. The increase is truly remarkable, as the prices between 1984–5 and 1991–2 more than doubled (from less than 2 million lire to over 4 million lire per m^2). Figure 5.1 shows a strong division between the central areas of the major Italian cities: the values of the central areas of Turin, Genoa and Naples have nearly halved in respect to those of Milan and Rome in the same time period. Between 1993 and 1995 the price trend line of Rome and Milan declined as fast as it climbed previously. Figure 5.2 shows a similar trend for new housing in peripheral areas of Rome and Milan. House prices in Genoa, Naples and Turin followed a comparable pattern, but their ups and downs are much flatter. This indicates, once more, that in Rome and Milan the housing

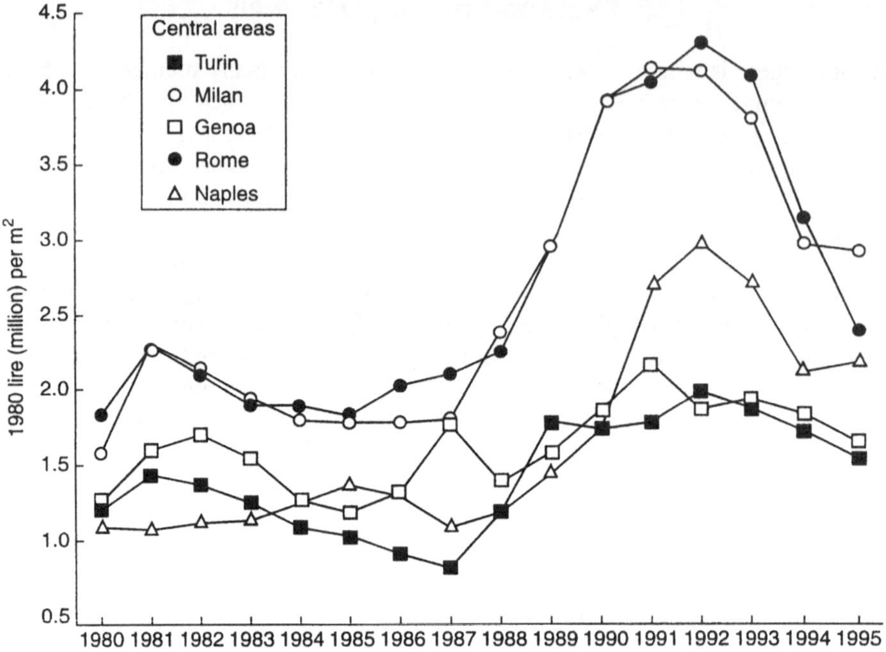

Figure 5.1 Average new housing prices in five leading cities in metropolitan areas (central areas).

market is more speculative than in the other major cities: the price gains can be greater then elsewhere in the country, but by the same token the price losses can also be massive.

A consequence of these different trends is that financial investment in the housing markets of Rome and Milan have yields whose trends are also subject to great variations, whereas in the other cities (notably in Turin) the stable course of housing prices generates a more predictable trend for yields. Stability seems to be a rather significant market feature from the viewpoint of institutional investors such as pension funds and insurance companies. Thus, a detailed knowledge of the different property markets available in Italy might eventually lead to a greater degree of interest in the housing markets of the largest Italian cities other than Rome and Milan. It is true that, until recently, the attention of investors in Italy was restricted to Rome and even more to Milan. This attitude can be explained by the fact that what was sought was a rapid appreciation of capital values of property assets, rather than good yields. In this respect, the property market of Milan was obviously the most interesting, but it also turned out to be the most unstable.

Indeed, the very idea that property investments should be valued according to their current and expected yields, and not on speculative gains, is basically "foreign" to the current expertise of the great majority of Italian property agents. This can be explained by the traditional role of "inflation shelter" that the property

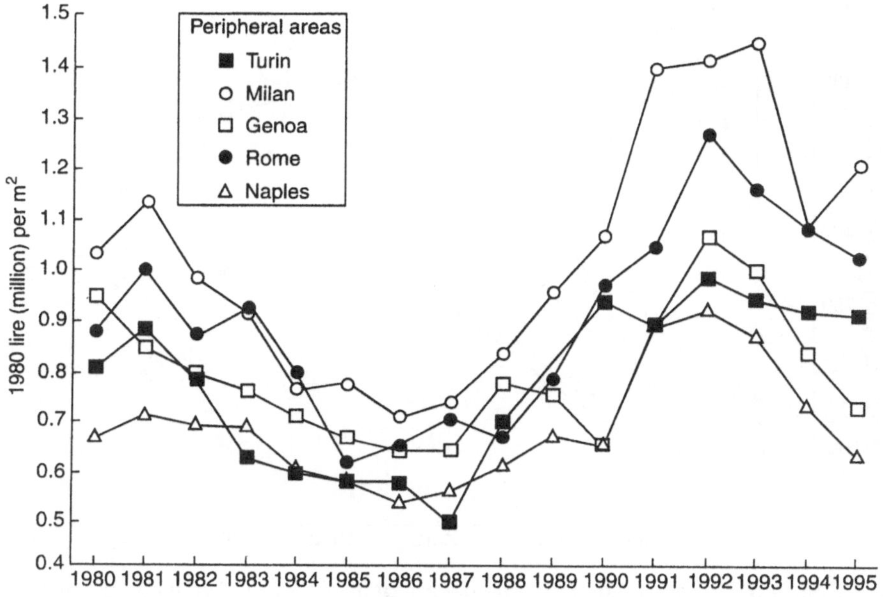

Figure 5.2 Average new housing prices in five leading cities in metropolitan areas (peripheral areas).

sector played in a country whose rates of inflation were in the range of 15–20% per year in the 1970s and until 1983 (see Table 1.4). However, this attitude should be abandoned. High inflation rates are not likely to return, whereas the market is changing rapidly and is becoming much more similar to the markets of other major Western countries. The housing market is now led by demand and not by supply. In addition, the fiscal pressure on the real estate sector is now a factor that can have a sizeable influence on the returns from property investments. Moreover, investors are traditionally more attracted to properties other than housing in a country such as Italy, where the yields in the rental sector have historically been capped for social reasons. Tenants could remain in rented housing for long periods (measurable in years, not months) in spite of legal procedures by the owners to obtain their eviction. Thus, empty housing units acquired a premium (about 30%) on the market price to mark their immediate availability as opposed to the "occupied" (rented) units. The same rationale was also applied to office units.

Figure 5.3 reports the actual selling price of a semi-central flat ($90m^2$) of medium quality for each of the five largest Italian cities and, an average, for the thirteen major Italian cities, for the period 1988–95. In 1988 the selling price in the five cities was in a limited range between 140 to 180 million lire, with little variance around the medium value. In 1991, at the top of the growth period, the average price of a standard flat in Milan had reached 420 million lire, about double the average price in Genoa. In Rome the value was 350 million lire. In sum, in 1991 the average price of a flat in Milan and Rome had reached a value

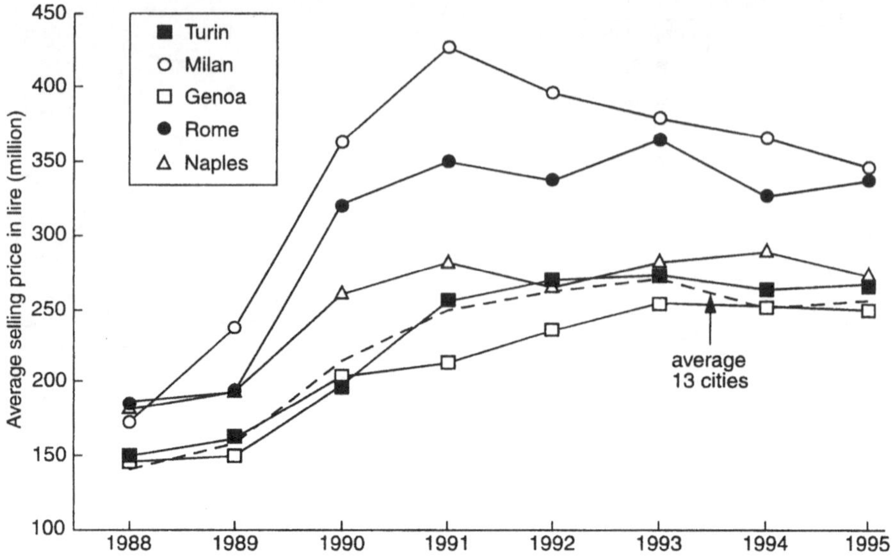

Figure 5.3 Selling price for a semi-central flat (90m^2) in five leading cities.

significantly higher than any other city in Italy. This became a serious hindrance to an efficient mobile labour market, because moving from the rest of the country to Milan or Rome would have implied a housing problem. From a macro-economic viewpoint, it seems clear that there is an inefficient match between the availability of jobs and the geographical distribution of employees and professionals whenever the rationale choice is influenced by external factors such as the rigidity of the housing sector. Obviously, one solution could be the existence of a flexible rental market. In Italy the rental market has been revived with the *Patti in deroga*, a variance to the house rent contracts enacted in the early 1990s to induce property owners to put their empty properties on the market. The measure has reintroduced into Italy the free negotiation of house rents. Nowadays the actual supply is greater then the demand, and rents have been declining after a period (1991–3) of increase. However, what has probably played a decisive role has been the introduction in 1993 of *ICI*, the tax on immovable properties, which has made empty properties a non-sustainable burden for the owners.

Table 5.2 describes the main characteristics of a sample of loans issued for the purchase of housing units in the 13 largest Italian cities. The data shows that, as of May 1995, the average housing value was 254 million lire, and the average mortgage loan was just 106 million lire, or 42% of the property value. This is a very low percentage, when compared to the average loans issued in other European countries or in the USA. In Italy the banking system has very conservative lending policies and an aversion to any kind of business risk. Even in the case of small and medium-size enterprises, lending must be backed by real estate assets. So, the decline of property values in current monetary terms, initiated in the first

Table 5.2 Main characteristics of a sample of mortgage loans issued for the purchase of housing units in the 13 largest Italian cities, May 1995 (millions of lire).

Cities	Average housing value (A)	Average mortgage loan (B)	% ratio loan (B) over value (A)	Average floor surface m^2	Price per m^2	Average year of construction
Bari	280	126	45.0	103.4	2.6	1978
Bologna	242	92	38.0	101.5	2.4	1964
Cagliari	221	108	48.9	3111.0	1.9	1974
Catania	172	67	39.0	105.2	1.6	1967
Florence	336	129	28.4	111.5	2.9	1963
Genoa	243	120	49.4	106.3	2.4	1962
Milan	301	126	41.9	97.9	3.0	1964
Naples	255	109	42.7	104.1	2.5	1960
Padua	258	102	39.5	145.5	1.8	1980
Palermo	249	107	43.0	132.9	1.9	1969
Rome	287	105	36.6	96.6	3.0	1970
Turin	227	83	36.6	105.4	2.2	1967
Venice	236	106	44.9	111.7	2.1	1971
Average	254	106	41.8	110.2	2.3	1968

Source: *Nomisma, Osservatorio sul mercato immobiliare* **2**, 1995.

half of the 1990s, has led the banks to be even more cautious in their lending to the real estate sector. The table also shows that the acquired housing units have an average floor-space of about 110 m^2. In no city is the average space less than 100 m^2, with the exception of Rome and Milan, where the housing units bought with a mortgage loan are on average just around 97 m^2. Apparently, high price levels impose lower consumption of space on households.

The decline of the market values of housing units is clarified by a close analysis of the selling prices at current values of housing units by housing condition in the 13 largest cities, from 1988 to 1995. As indicated by Table 5.3, the turning point was in 1992 and 1993. In the case of new or completely renovated housing units, the average price rises constantly from 1.6 million per m^2 in 1988 to 3.1 million in 1992 and 3.2 million in 1993, the peak value of the period. In 1994 the average price decreased to 3.0 million per m^2, with a modest rise in 1995.

The data in Table 5.3 also shows the great differences in market values in Venice. Within each of the three categories of housing condition, the housing units of the historical Venice (on lagoon) are valued at approximately double per m^2 of those located on the firm land of the commune.

The analysis of prices of retail properties in the five leading cities of metropolitan areas shows a general trend similar to that of housing property (Fig. 5.4). It should be noted that the price range is much wider between cities than in the housing market. Again, the top prices are found in Rome and Milan, where the average price for a retail unit reached the record high values of 6.0–7.0 million lire per m^2 between 1991 and 1993. In the same period the average price in Turin, Genoa and Naples was in the range of 3.0–4.0 million lire per m^2 (with an unusual value above 5.0 million lire in Naples in 1991). A comparison between peripheral

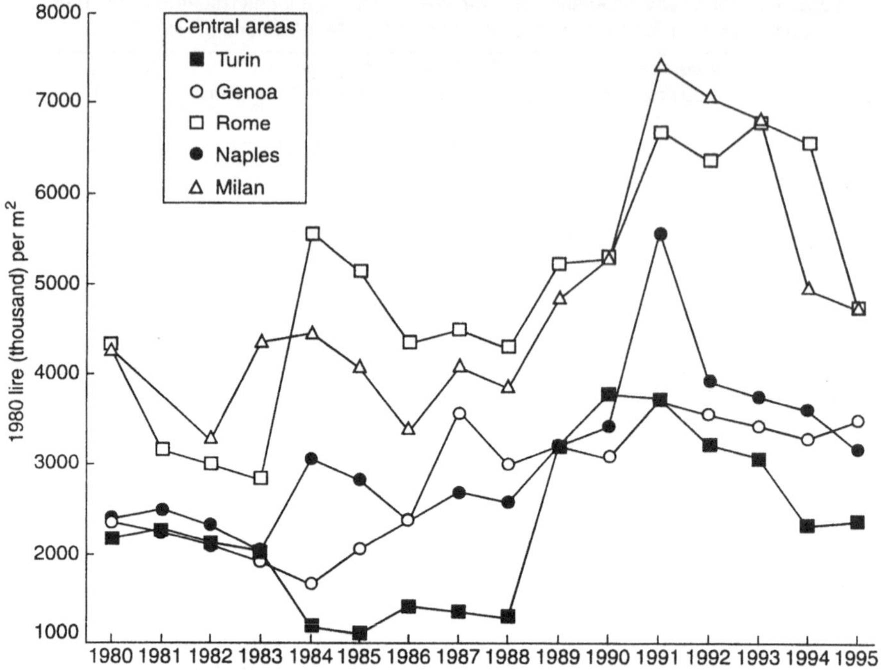

Figure 5.4 Average prices of retail properties in five leading cities (central areas).

and central areas indicates that there is a ratio of about 1 to 3 between the average prices of retail properties (Figs 5.4 and 5.5). For example, in 1980 the average price of retail property was 1.4 million lire per m^2 in the peripheral zones of Rome and about 4.3 million lire in central areas. The variance would be even greater in cases of a direct comparison between prime locations (which are the best central locations) and the peripheral areas. Figures 5.4 and 5.5 highlight:

- a period of crisis in the years 1981–4, with the exception of the markets in the central zones of Rome and Naples, which showed a rise between 1983 and 1984
- a period between 1984 and 1988 characterized by slumps in the peripheral zones of the five cities and paths much different from those of the central zones
- a general growth of the values in recent years, followed by a decline in the central areas of Rome, Milan, and Naples. In Turin and Genoa the average price shows modest increases between 1994 and 1995. This might be evidence of the effectiveness of some local policies directed towards restoring the central parts of these two cities.

The effects of the recent economic depression are evident in the trends in retail property prices in the other five large cities (Venice, Florence, Bari, Bologna and Palermo). There was a relatively homogeneous tendency between the cities in the period preceding 1983 and a situation very different according to the cities

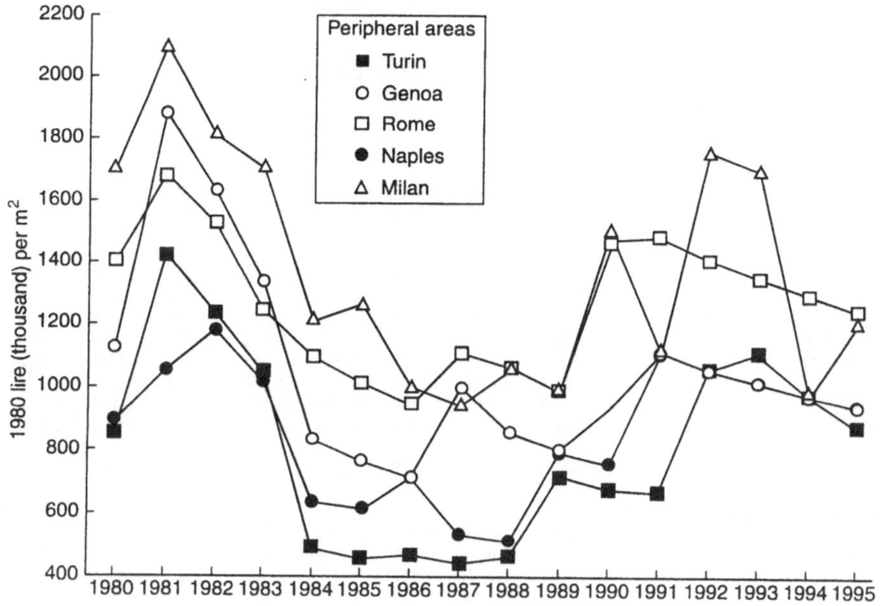

Figure 5.5 Average prices of retail properties in five leading cities (peripheral areas).

and the location (central or peripheral areas) in the following years.

The analysis of price trends for the retail and office sector can now be focused on the general trends which can be observed in Italy. Figures 5.6 & 5.7 have been drawn from the data available in the period 1988–95 for both retail and office properties in the 13 largest Italian cities covering all urban areas (prime location, central and medium centre, peripheral areas) of each city. The resulting trends are rather similar: in both cases the values observed in Rome and Milan are on absolute levels much higher than the average. From 1988 to 1992 (Fig. 5.7) there is a growth in the average values of retail properties in the 13 cities from 2.2 million to almost 4.0 million lire per m^2, whereas in the subsequent period 1992–5 there is a constant decline, with the 1995 figure being around 3.3 million per m^2. The price trend for Milan shows an increase from 2.5 million to almost 6.5 million lire per m^2 in the period 1988–91, followed by a decline to 4.5 million in 1995. In the case of Rome the peak value is reached in 1992, with over 7.0 million lire per m^2, and the decline is sheer down to 4.6 million per m^2 in 1995. Thus, at the end of the period, the price values of retail properties in the five largest Italian cities fall within a range of between 2.5 million (Turin) and 4.6 million lire (Rome), that is, in a range as wide as at the beginning of the period, whereas the lowest value was 1.8 million (Turin) and the top value 3.3 million (Rome).

In the case of office properties (Fig. 5.6), the average value per m^2 moved from 1.6 million lire to about 2.8 million in 1993, and to 2.5 million in 1995. The average price in Milan reached a maximum value in 1991 (4.5 million per m^2) and declined in the following years, remaining above the values of Rome, except in

Table 5.3 Weighted average selling prices of housing by housing conditions in the 13 largest cities, from 1988–95, in lire (million) per m^2.

	1988	1989	1990	1991	1992	1993	1994	1995
New or completely renovated housing								
Bari	1.092	2.148	2.283	2.673	2.919	2.836	2.637	2.602
Bologna	1.970	2.403	3.036	3.431	3.996	4.023	3.721	3.782
Cagliari	1.488	1.608	1.747	1.888	2.252	2.366	2.443	2.458
Catania	1.170	1.353	1.514	1.568	1.782	1.799	1.733	1.950
Florence	1.768	2.277	3.079	3.581	3.749	3.804	3.686	3.724
Genoa	1.727	2.313	2.505	2.594	2.866	2.863	2.868	2.877
Milan	2.001	2.891	4.574	5.151	5.016	4.909	4.486	4.575
Naples	1.715	2.052	2.523	2.751	2.835	2.922	2.818	2.786
Padua	1.262	1.554	1.844	2.317	2.613	2.799	2.600	2.801
Palermo	1.097	1.463	1.811	2.208	2.492	2.468	2.130	2.080
Rome	2.100	2.996	3.089	4.092	4.267	4.191	3.797	3.843
Turin	1.638	1.992	2.504	2.947	3.248	3.204	3.073	3.084
Venice on lagoon	1.913	2.781	3.001	3.462	3.663	4.037	4.069	4.179
Venice on ground	1.106	1.444	1.867	2.141	2.255	2.501	2.524	2.542
Average	1.633	2.091	2.578	2.915	3.140	3.194	3.042	3.092
Old housing in good condition								
Bari	1.382	1.535	1.953	2.080	2.218	2.170	2.152	2.189
Bologna	1.371	1.818	2.315	2.704	3.105	3.067	3.023	2.994
Cagliari	0.947	1.247	1.306	1.392	1.678	1.877	1.946	2.024
Catania	0.964	1.100	1.214	1.287	1.477	1.556	1.519	1.713
Florence	1.517	1.874	2.640	3.064	3.297	3.387	3.177	3.161
Genoa	1.362	1.640	1.849	1.993	2.216	2.228	2.200	2.132
Milan	1.690	2.342	3.760	4.314	4.110	4.029	3.731	3.704
Naples	1.483	1.608	2.192	2.377	2.471	2.408	2.363	2.346
Padua	1.015	1.255	1.510	1.852	2.055	2.016	2.117	2.160
Palermo	0.870	1.128	1.395	1.685	1.874	1.796	1.658	1.578
Rome	1.786	2.566	3.242	3.473	3.578	3.848	3.445	3.202
Turin	1.175	1.485	1.852	2.297	2.526	2.597	2.390	2.478
Venice on lagoon	1.565	2.542	2.455	2.746	3.036	3.363	3.471	3.469
Venice on ground	0.905	1.073	1.413	1.538	1.737	1.876	1.939	1.965
Average	1.288	1.658	2.078	2.343	2.527	2.598	2.509	2.508
Housing requiring complete renovation								
Bari	1.032	1.166	1.515	1.554	1.619	1.629	1.725	1.665
Bologna	0.973	1.294	1.723	2.031	2.277	2.309	2.427	2.307
Cagliari	0.699	0.780	0.885	1.041	1.166	1.394	1.498	1.638
Catania	0.675	0.832	0.988	0.994	1.148	1.165	1.215	1.361
Florence	1.470	1.497	1.918	2.275	2.647	2.847	2.662	2.588
Genoa	1.004	1.091	1.304	1.143	1.631	1.719	1.611	1.620
Milan	1.266	1.920	2.729	3.174	3.393	3.325	3.044	2.976
Naples	1.120	1.217	1.609	1.741	1.878	1.959	1.955	1.894
Padua	0.541	0.816	1.118	1.254	1.457	1.445	1.388	1.436
Palermo	0.599	0.723	0.951	1.167	1.279	1.347	1.224	1.211
Rome	1.474	2.085	2.653	2.821	3.024	3.231	2.944	2.498
Turin	0.838	1.302	1.298	1.616	1.841	1.938	1.725	1.814
Venice on lagoon	1.150	1.841	1.853	2.166	2.446	2.646	2.813	2.807
Venice on ground	0.548	0.794	1.084	1.299	1.343	1.476	1.396	1.483
Average	0.956	1.221	1.545	1.753	1.939	2.031	1.973	1.950

Source: *Nomisma*, *Osservatorio sul mercato immobiliare* **2**, 1995.

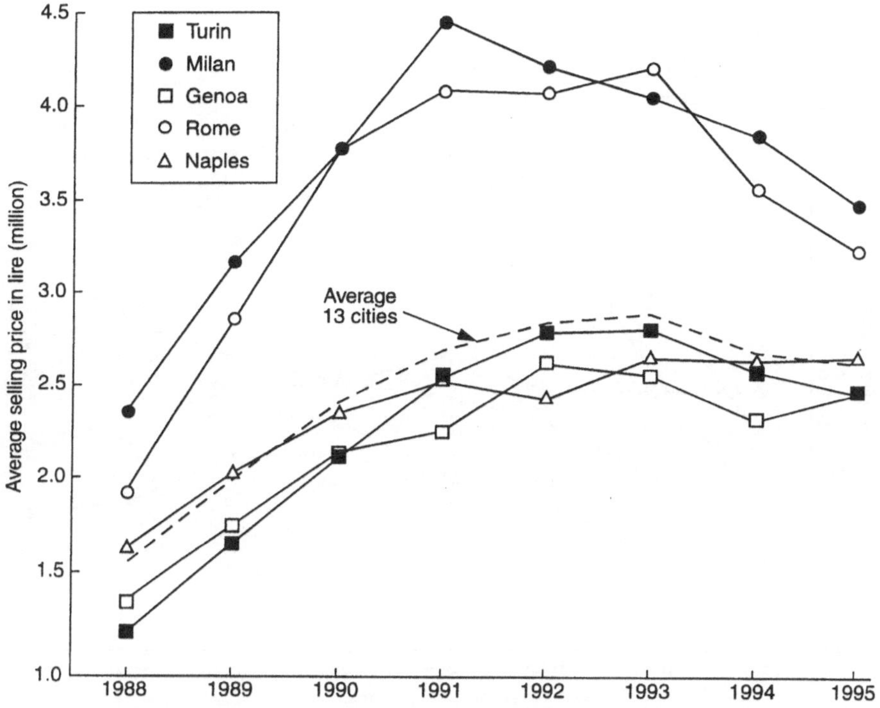

Figure 5.6 Price of office properties in five leading cities.

1993 when the office market of the Italian capital recorded its top average value of about 4.3 million lire per m^2.

The data provided in Figures 5.6 & 5.7 indicate price trends that have generated a real shock within the Italian real estate community. In fact, the Italian market has historically been based on expectations of a continuous capital appreciation of property assets. The general training of most real estate experts was rooted in the housing sector, where there was the unquestioned assumption that property assets were the best shield against the devaluation of the Italian lira and so the market values of property assets could only increase over time. The introduction of *ICI* (the tax on immovable properties) in 1993 has brought a substantial weight of taxation to property assets and, coupled with the decline of current property prices, it may well be that the increased fiscal pressure in the property sector will finally persuade the Italian property community that a perennial increase in property values is a thing of the past. Eventually, the analysis of property yields as it is currently undertaken in other countries will also make its way into the Italian market. Property assets, other than housing for own use, will be analyzed and judged primarily for their expected net yields and not for their supposed role against monetary depreciation or for speculation.

Tables 5.4 and 5.5 report the average annual yields of housing, retail and offices in the 13 largest Italian cities, as of May 1995. With regard to the housing

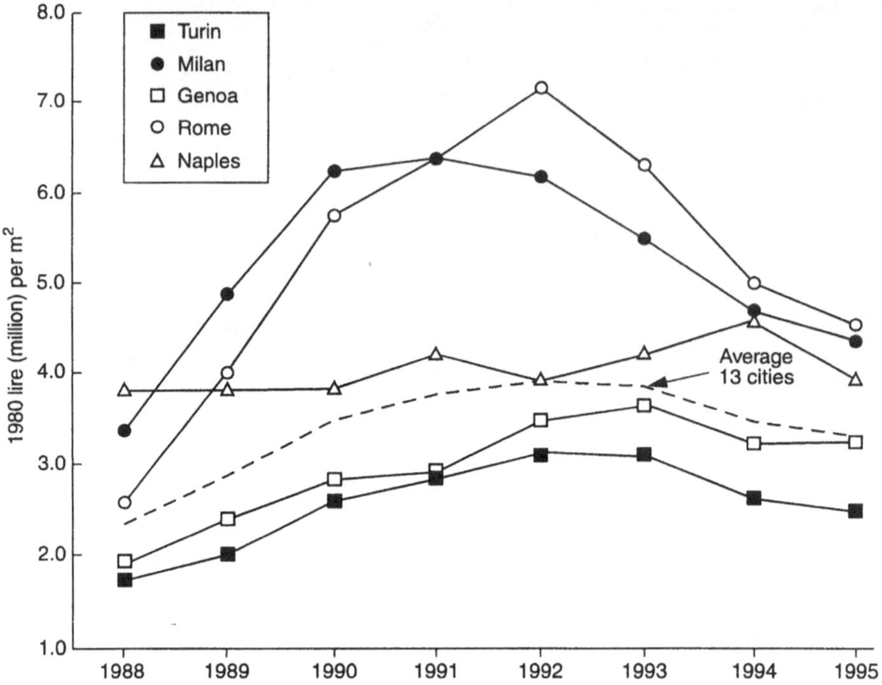

Figure 5.7 Price of retail properties in five leading cities.

market, it may be noted that the average yields exhibit a very regular rising pattern from the prime areas (5.12%), to central zones (5.18%), semi-central locations (5.48%) and peripheral locations (5.73%). This is consistent with the distribution of housing market prices which, on average, have an opposite trend, declining from prime and central areas towards the outskirts of the analyzed cities. A share of the rented housing sector is still regulated by the Rent Control Act, which on the one hand sets price caps on the amount of rent but, on the other hand, has the effect of ensuring a moderate growth of yields as the rent is raised automatically every year at a rate of 75% of the official consumer price index (CPI) in Italy. This means that rents are rather rigidly fixed and do not follow the fluctuations in the market prices for rented housing. The over supply of housing has had the effect of depressing the market prices of housing in areas of no special character, such as the peripheral zones. These factors have led to lower market values for rented homes, while rents did not decline, especially in the periphery where they were already at lower levels compared to other urban areas. Hence, housing yields are generally higher in the outskirts compared to the central zones.

Concerning retail properties, the observed average yields appear to be constant throughout the urban locations: 7.69% in prime commercial areas, 7.59% in semi-central areas, and 7.54% in peripheral zones. In general, the yields are higher than the rental housing sector. A closer look reveals that retail properties

Table 5.4 Average annual yields (%) of housing in the 13 largest cities, by location in the urban area, May 1995.

Housing rented in free market	Prime area	Central area	Semi-central area	Periphery
Bari	4.40	4.40	4.75	4.45
Bologna	4.50	4.75	4.50	4.30
Cagliari	6.35	6.05	6.25	6.10
Catania	4.95	5.70	5.45	6.35
Florence	6.40	6.35	6.20	6.05
Genoa	4.25	4.75	5.15	6.30
Milan	4.25	4.20	4.75	4.35
Naples	4.55	5.15	6.60	7.25
Padua	5.35	4.95	5.90	7.05
Palermo	5.40	5.15	5.75	6.70
Rome	6.95	6.25	6.60	6.90
Turin	4.05	4.05	4.25	4.50
Venice on lagoon	5.10	5.60	5.30	4.80
Venice on ground	5.15	5.20	5.20	5.15
Average	5.12	5.18	5.48	5.73

Source: Nomisma, Osservatorio sul mercato immobiliare **2**, 1995.

in the periphery have higher yields than those in prime areas in the majority (7 out of 13) of the analyzed cities. The average yield for the 13 cities is heavily influenced by Venice, where the average annual yield is extremely high (over 11%) both in the historic city on the lagoon and in the modern city (Mestre). There is now evidence that both the market values and the average yields of retail properties are declining, under the pressure of new shopping centres, which are appearing in great numbers, especially in the Centre and North of the country.

In the office market the analysis is focused on four, not three, standard locations within the urban area. In fact, besides the usual subdivision of centre, semi-centre and peripheral areas, there is the new phenomenon of office clusters which have peripheral locations but provide quality buildings and services that compete with the office supply of the whole urban area. It can be observed that the yields for office properties exhibit a straight line rising from the prime and central locations towards the periphery and the office clusters, whose location is often beyond the administrative boundaries of the large communes. The average yields as of May 1995 are as follows: 5.99% in prime locations, 6.13% in semi-central areas, 6.41% on the periphery and 6.48% in the office clusters. As in the retail and housing markets, it may be noted that in some cities the office properties have yields higher than those in the cities of Rome and Milan.

Milan and Rome have a high visibility for foreign investors, but they do not appear to offer the highest yields nor the most stable ones, when compared to the other largest Italian cities. This is true for all the analyzed markets (housing, retail and office properties). One may wonder if the persistent attention of national and foreign investors to these two cities is a rational choice or, as is most likely, the result of lack of knowledge of feasible investment alternatives.

Table 5.5 Average annual yields (%) of retail properties and offices in the 13 largest cities, by location in the urban area, May 1995.

	Prime area	Semi-central area	Periphery	Office clusters
Located retail properties				
Bari	5.75	6.85	6.05	
Bologna	6.45	6.75	7.50	
Cagliari	9.25	8.95	7.95	
Catania	7.80	7385	8.40	
Florence	6.00	6.75	6.00	
Genoa	7.85	6.80	7.10	
Milan	7.35	8.10	8.35	
Naples	6.50	6.85	7.90	
Padua	9.25	8.20	8.35	
Palermo	6.10	6.50	7.70	
Rome	6.80	7.55	6.80	
Turin	6.35	6.85	6.45	
Venice on lagoon	11.10	8.45	8.80	
Venice on ground	11.05	9.75	8.25	
Average	7.69	7.59	7.54	
New or completely renovated offices				
Bari	6.60	6.40	6.25	7.15
Bologna	4.90	4.85	4.95	5.35
Cagliari	6.50	6.50	7.55	7.00
Catania	6.25	6.45	6.50	7.55
Florence	6.55	6.55	6.35	6.15
Genoa	5.25	5.35	5.90	5.95
Milan	5.90	5.75	6.05	5.90
Naples	5.95	6.10	7.75	6.95
Padua	6.00	6.25	6.65	6.95
Palermo	5.30	6.15	5.85	5.85
Rome	6.45	6.35	7.65	7.05
Turin	5.00	5.15	5.15	5.15
Venice on lagoon	5.70	6.35	5.85	
Venice on ground	7.55	7.55	7.35	7.20
Average	5.99	6.13	6.41	6.48

Source: *Nomisma, Osservatorio sul mercato immobiliare* **2**, 1995.

5.4 Speculation in property

Speculation mostly concerns the rental market, the transformation of a property's use and the facilitated housing market. The market for rented housing has been regulated in Italy by the 1978 Fair Rent Act. This defined an administrative price system which set the rent equal to 3.85% of the conventional value of the property, which is estimated according to a number of parameters, and tied in to the length of the contract.

There is a general view that the relationship between the interests of the owner and the tenant was inadequate, in terms of both title of ownership and of its use, and in terms of the economic value placed on the lodging. In fact in many areas of the country the law sanctioned local values superior to those that would have been determined by supply and demand, whereas in a few areas (especially metropolitan and tourist areas) the law retained the values that would have been determined in the market. As a result of this there has been: a contraction of the rental market, with a consequent formation of ample empty stock, especially in the large cities; a rise in the demand for property, which has contributed to the rise in housing prices; and speculative activity in the renting market.

In the early 1990s, however, the rental market has been revived by the introduction of *Patti in deroga*, that is, rent contracts with no price cap on the legal amount to be charged to the tenant, in the framework of contracts with longer validity periods. The situation is now completely changed in the private rental market, with an over-supply of properties for rent and consequently falling prices.

Another type of speculative phenomenon in the 1980s was associated with the change in use of the property from residential to the service sector, a widespread phenomenon in the centres of metropolitan cities. This guarantees free rent negotiation, whereas the actual bonds on the length of the contract are kept the same. In the mid-1990s, with the well known over-supply of office properties, the reverse is happening. Some owners of unsold and unlet offices try to convert their properties into housing without the required approval from the local authorities. So far, there have been only a few cases of this, mainly in Milan, where the bulk of empty offices can be found.

There is a third type of phenomenon, especially in the surroundings of the metropolitan areas in which the recent construction of assisted residential buildings has occurred. For such properties, access involves accepting the law which permits the possibility of sale only after five years and at conventional prices. This minimum time period is sometimes eluded, with the anticipated sale of properties at near-market values and with the consequent windfall gain. This involves very few households, and it does not have major effects on the outcome of the assisted housing construction programmes.

5.5 Demand and supply of land and property

Demand and supply of building land

The market of building land in Italy is very difficult to analyze, because of the paucity of available information concerning the number of transactions of land plots and the actual prices paid. The cause of this is the fact that, in the legal contract, both seller and buyer tend to declare a lower value in order to pay less taxes. The seller would pay a lower capital gains tax, whereas the buyer would avoid having his income potential uncovered by the full amount paid for the property. However, this practice is tending to diminish since the introduction of *Imposta comunale sugli immobili* (ICI, the local tax on real estates), which requires an annual payment linked to the cadastral value of the real estate assets. This has made the cross-checking of the yearly income declaration of individuals and of the balance sheets of firms rather easy. The cross-checking is very severe in a case where taxpayers have been expropriated of their real estate assets and when they are further the object of a detailed review of their income declaration at the request of the Ministry of Finance. In this case, taxpayers have to refund any excess money received as settlement for the loss of property, over and above the declared value of the same property for tax purposes in the year before the expropriation. These are rare cases, but as they are widely publicized by the media they tend to act as hindrance to cheating for the vast majority of taxpayers who own properties.

The marked differences between the metropolitan areas and those of the minor urban centres, between the land use (residential, commercial, industrial and services) and between regimes of a public, private or even of a mixed nature, are among many factors that characterize such a complex land market. In the largest cities and in fashionable tourist areas, developable land is rare, as in many cases development is halted because of planning restrictions, sometimes coupled with a scarcity of land suitable for construction. The price range of land is very extended (from just a few thousand lire to millions of lire per m^2). Finally, the question of dispossession of land for public use is more than a hundred years old in Italy, and legislation is very complex (see §3.1). This issue has to do with the interests and rules of economic life, such as the right of ownership, which constitutes a critical role that has been and remains up until now discussed on a national level. While legislation now covers all phases of the expropriation process, there is not yet a comprehensive land act that would simplify its application. This is a task which is likely to be on the agenda of the national governments in the next decade.

It is now possible to describe the course and the consistency of the percentage incidence of the costs of land on average property values in the central and peripheral zones. The data have been gathered with reference to the housing market in the five largest metropolitan centres and in another five important Italian cities (Bari, Bologna, Palermo, Florence and Venice).

For the latter (Fig. 5.8), the incidence of land value in peripheral zones is within the range of 10 percentage points throughout the 1980–95 period (e.g. it is between 30% and 40% in 1995), with the exception of Palermo, which registered the lowest prices in 1991 and 1992 (20%). It is noteworthy that the values of peripheral areas in the largest cities (Fig. 5.9) are in line with the smaller ones. The value of the land comprises, throughout the 1980–95 period, between 35% and 50% of the property price, with a peak of 55% in Milan in the years 1991 and 1992. The latter may have been caused by rumours of a relocation of the Milan Trade Fair and other important facilities to a site outside the city. Also, in Florence there is an equivalent rise

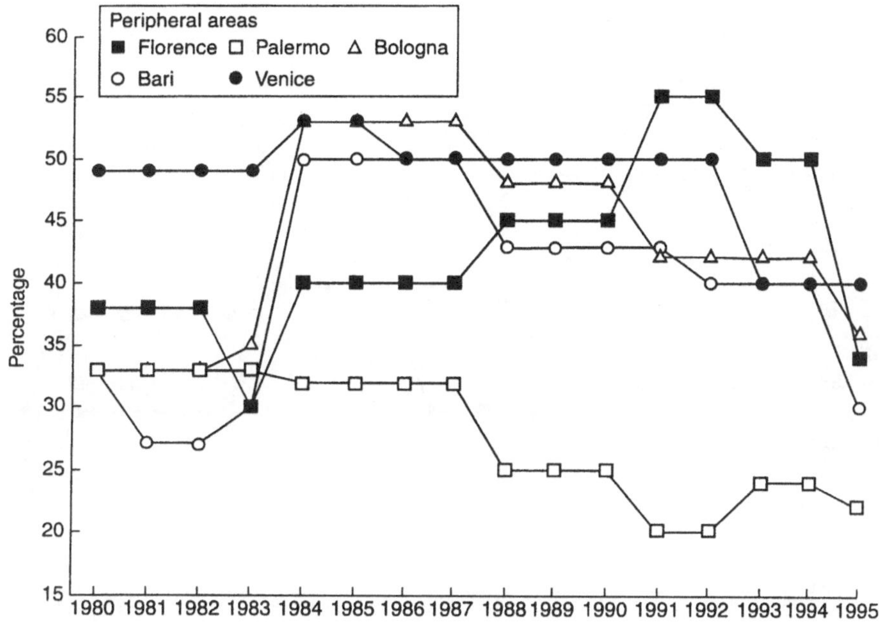

Figure 5.8 Percentage incidence of the area on the price of property in five major Italian cities (peripheral areas).

from 45% to 55% of the land weight in peripheral property prices. In this case, the rise may have some connection with the large-scale development project that was being implemented in Novoli, on the outskirts of the city.

The trends may suggest that the market evaluates the land factor of the peripheral areas of large cities at the same level of equivalent land plots in smaller cities. The outskirts of a small city are, after all, much closer to the respective city centres than the vast fringe areas of Rome, Milan or Turin. Hence, the property values of peripheral areas of smaller cities are closer to the top values than the equivalent properties in the largest cities. Most of the new housing and office development of the first half of the 1980s took place on the fringes of the largest cities. This made the land factor rather important and enhanced its share of the total property value. In the second half of the decade, as it was shown by the illustrations in §5.3, there was a sharp rise in the market prices of housing and offices located in prime locations, which with few exceptions coincide with the central areas. This increase had an evident pulling effect on the whole structure of property prices in the large metropolitan areas, until the beginning of the current real estate slump in 1990–91.

The central locations of both clusters of cities (Figs 5.10 & 5.11) exhibit a rather comparable price trend, with two rising waves of values at the beginning and at the end of the 1980s. The peak values at the centres of Rome and Milan are particularly high (70%–80% on the sale price in 1990), whereas in the second

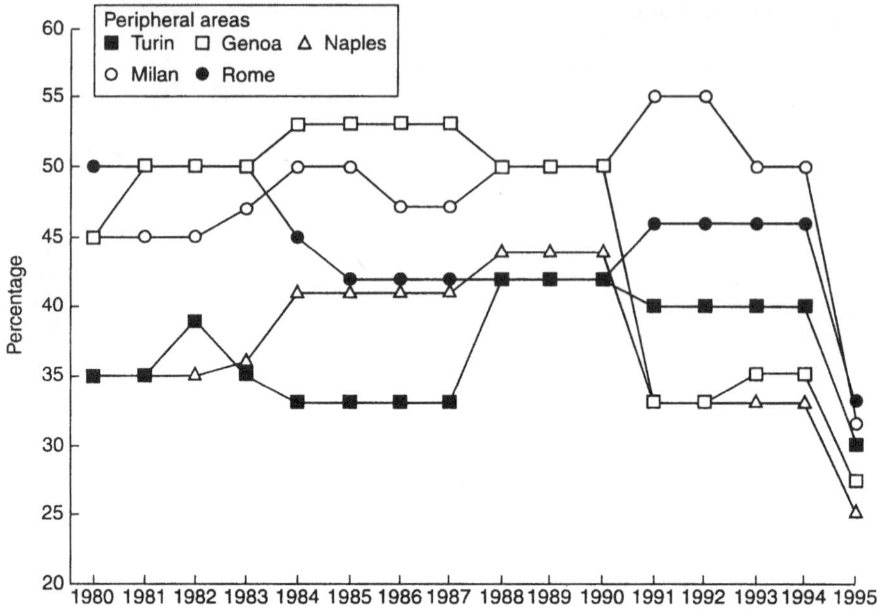

Figure 5.9 Percentage incidence of the area on the price of property in five major cities (peripheral areas).

cluster of analyzed cities the situation seems to be more comparable. A closer look shows that the price trend of the largest cities (Fig. 5.11) has anticipated the trends of the other group. In fact, the incidence of the land increases in Rome from 70% to 80% of the property value between 1982 and 1985. In the same period there is a steep increase in Naples, from 55% to 70%, whereas in Milan, Genoa and Turin there are less marked increases. In the other cluster (Fig. 5.10) the percentage incidence of land starts the upwards trend only in 1983, that is, one year later than in the largest cities.

In the first half of the 1990s there is a decline of the weight of the land factor. It is very accentuated between 1994 and 1995. The incidence goes from 80% to 65% in Naples, probably as a result of the marketing of the new Naples Business District, which had a depressing effect on the values of the land for other office projects. The decrease is also marked in Turin and Genoa, where the incidence of the land value on total property price moves respectively from about 67% to 52% and from 65% to 50%. In Milan the decline is less accentuated, from 75% to 70%.

In all the three cities, which had been regarded in the past as the vertices of the Italian industrial triangle, a possible explanation is related to the huge vacant industrial areas made available for new uses. This additional supply had negative effects on the absolute values of urban land, as it is now obvious that the market is unlikely to sustain all the development projects that could be built on the available

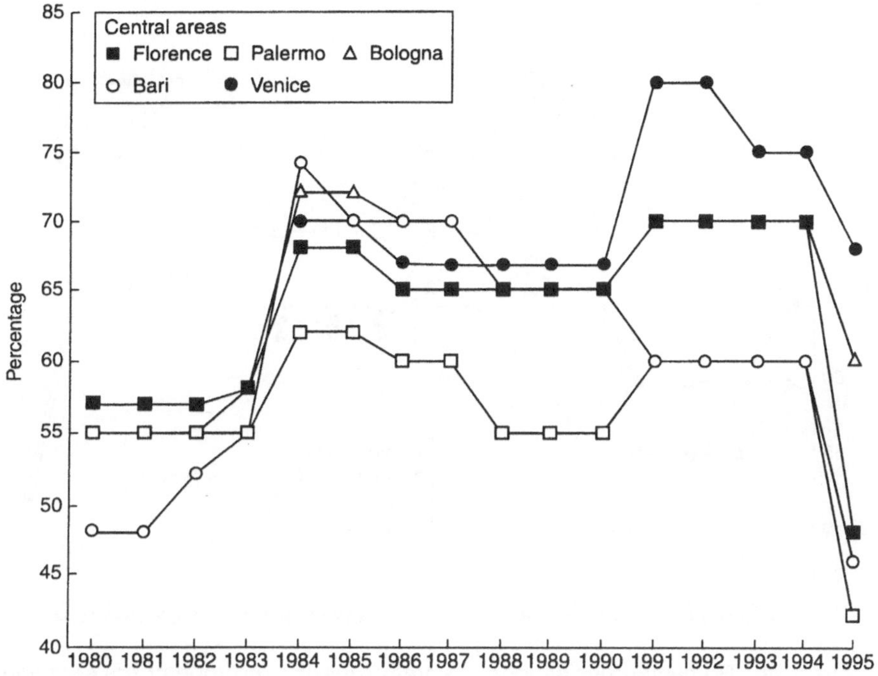

Figure 5.10 Percentage incidence of the area on the price of property in five major Italian cities (central areas).

plots. The problem requires a coordination between public and private forces, as local governments would also be negatively affected by the possible failures of private development projects. It would seem appropriate to consider a kind of phasing-out strategy, so that the vacant industrial land would not reach the market all at once.

Only in Rome does the incidence of the land factor remain stable, at around 70% until the end of 1995. The property market of the capital has always had unique characteristics. Compared to other areas, Rome is the stage of recurrent special events because of its role as "Eternal City". Besides the presence of ministries and international institutions, the widely publicized programmes of public works for *Roma Capitale* (Rome capital) and for the SDO (*Sistema direzionale orientale*, the transferral of key ministries from the city centre to a vast area on the east side of the city) or for *Giubileo 2000*, have created expectations that may be reflected in current high levels of land prices, which anticipate demand even before any of the mentioned programmes are implemented.

Unused industrial land

At present, there are no reliable sources on the phenomenon of vacant industrial land in Italy. According to a survey conducted by the institute *Scenari Immobiliari*, and reported by *Sole 24 Ore* (the leading financial newspaper) in March

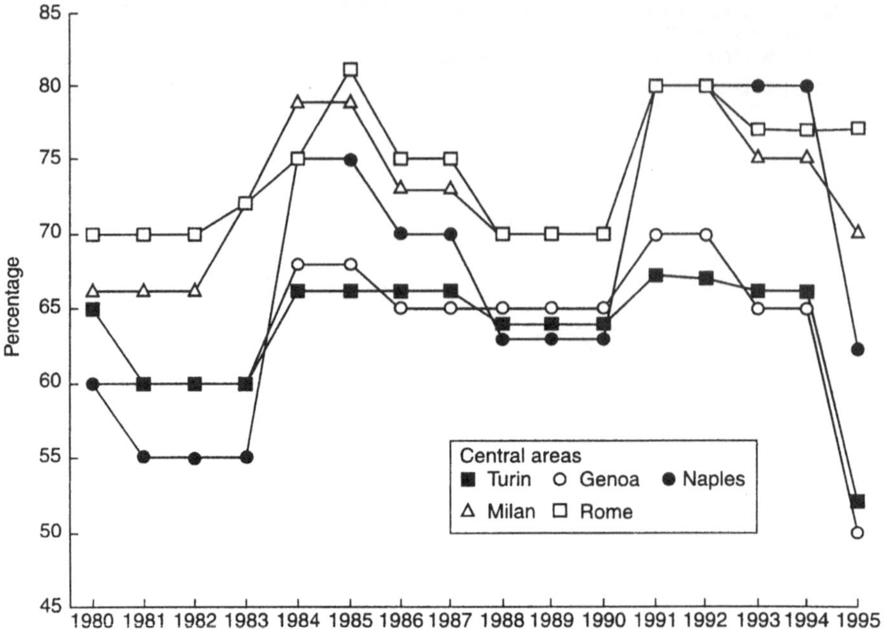

Figure 5.11 Percentage incidence of the area on the price of property in five leading cities (central areas).

1996, the total quantity of vacant land is 90–100 million m^2. This figure seems over-estimated. In fact, the same source reports the results of a detailed survey of vacant industrial land in the province of Ferrara, where out of a total of 9 million m^2 of vacant and decaying industrial properties, 5.2 million m^2 are located at the *Saline di Comacchio*, a flat area by the Adriatic Sea where salt has been produced since ancient times. By the same token, the abandoned mines of Sardinia should also be included in vacant industrial land. The truth is that by "vacant industrial land" the real estate market refers exclusively to the *urban* industrial land that has become available for new uses as a consequence of the transferral of industrial production. Quite simply, there are two kinds of vacant industrial properties: inside and outside urban centres. In the first case, the source of interest is not the existing use of the land, but the location of the land inside the city, often with excellent accessibility to the city centre and to natural facilities such as waterfronts. The positional rent accrued over the decades makes the former industrial land inside large cities highly appealing for property developers. As the real estate market has been in an over-supply phase since the late 1980s, only the vacant land with the best accessibility and in the most dynamic urban centres can generate a real interest among investors and end-users. In the second case, the re-use of vacant industrial land located outside urban centres is more a problem of restoring the natural environment than an issue for the property market.

In Milan, the availability of areas already vacated by installations or those free for new uses is in the order of six million m^2, with various plots exceeding 100000 m^2 and some of 500000 m^2. In the metropolitan area of Turin, the presence of industrial wasteland available immediately for conversion of use is above three million m^2. Plots vary from 20000 to more than 100000 m^2. Indeed, the re-use of vacant industrial land became the cornerstone of the city master plan approved in 1994. Basically, all new development envisaged by the plan is to take place in the former industrial land. This will also mark the transition of the economic base of the city from industry to new technologies and service activities.

In Genoa there are significant vacant areas throughout the city, and above all in the zone of the old port and of the River *Polcevera*. In Rome there are large vacant (or greatly under-used) areas, but only a few industrial properties, so the phenomenon is more complex and requires further analysis. Naples presents a situation of scarce concentration, with the exception of the plants of *ITALSIDER* of Bagnoli, whose complete re-use is high on the agenda of the new city administration. The site of Bagnoli has been contaminated by decades of steel production. To clean up the site the central government approved a budget of 250 billion lire in July 1996.

The cities just mentioned are only part of the complete picture of the "empty cities", whose land values ultimately depend upon new destinations of use, urban standards and land density. In sum, the value of the old industrial land is completely dependent upon the prescriptions of the city master plan.

In the whole country the industrial areas inside urban centres might not exceed a range of 20–25 million m^2. Obviously, if all these areas were to be developed at the same time, the market would be flooded by a new wave of unsold and unlet properties. The old industrial areas usually have fairly central locations, with excellent accessibility to the main infrastructures. In the intensively built Italian cities, these large and under-utilized properties have become crucial for urban redevelopment. Thus, the decision on what to do for these areas is often the result of a complex and long debate on the future of the whole city. It is felt that the future direction of development of the economic base of the city depends on the present decisions taken on the old industrial areas. The pressure on public administrators is very high. It is clear for them that the existence of abandoned areas is a social cost for the city. The landowners are also under pressure, because the unused assets have a negative impact on their annual budget. Yet, the problems are so complex and entail such large interests that the result is often a non-decision.

Nevertheless, it seems evident that in Italy most urban projects in the next few decades will be essentially located within the already built-up areas of large cities. New developments on greenfield locations are less and less socially sustainable, whereas the re-use of vacant properties is favoured by both city administrators and residents and obviously by the owners of the properties. The question is what to do with the old industrial land, given the conspicuous saturation of all real estate markets in the 1990s. In fact, industrial firms that relocated from urban to out-of-town areas in the 1960s and 1970s were often able to speculate on their property assets. The simple but effective bargaining attitude ensured that relocation was postponed until the city administration granted planning permission for housing development in the old area, which was considered a no-risk business.

In the 1980s there was a trend to build offices instead of housing, with the outcome that the market was flooded with so-called office centres located in fringe urban areas. Most of them were not a success.

As the unsold and unlet new properties became widespread in the late 1980s, private owners have increasingly recognized that what is needed to convert old industrial sites effectively is true cooperation with local government. To this end, several prominent Italian city administrations and private firms founded *AUDIS* (*Associazione delle aree urbane industriali dismesse*, the association of vacant industrial urban areas), whose mission is to promote economic efficiency and social equity in the re-use of industrial wasteland. The association, founded in Turin in 1995, includes as founding members the communes of Turin, Naples and Venice, and the following private firms and associations: ITALGAS (national company for gas provision), *Metropolis* (the real estate promotion firm of the state railway company *FS*), *FINSIDER–SECOSID* (the former state company for steel production), *SNAM* (the largest oil and gas producer of Italy), *ANCE* (the national association of construction firms). It is currently too early to evaluate the action of this association. However, its very existence seems to be a telling sign of a new way to approach the issue of development on former industrial land.

In conclusion, the prospects for the actual re-use of vacant industrial land rely once again on the kind of public–private relationships that will be implemented at the local level. Investors and end-users will probably direct their plans towards the cities where public administration (commune, province and region) presents an economic climate and a planning framework founded on sound cooperation between public bodies and with institutions such as the chamber of commerce, the associations of industrial enterprises and real estate developers, and private firms at large.

The re-use of the real estate assets of the Italian railway company.

The urban land owned by the *Ferrovie delle stato* (*FS*, the Italian railway company), is a special case of under-used industrial land within cities. Between the end of the 1980s and the early 1990s, the *FS* defined its strategy for the development of its real estate properties. In 1985 the *FS* became an *Ente Autonomo* (an independent institution) and in 1992 a *Società per Azioni* (a stock company with the same legal status as private corporations). By the end of 1991 the *FS* had produced the programme (called *Programma direttore nazionale per le grandi aree urbane*) for the period 1992–2001 concerning the restructuring of the group and the real estate diversification programme for the properties owned by *FS*. The key objective of this programme was to develop the real estate assets located in the first 14 cities planned to be served by the high-speed railway system of Italy: Turin, Milan, Verona, Venice, Genoa, Bologna, Florence, Rome, Naples, Salerno–Battipaglia, Bari, Reggio Calabria, Messina and Palermo. Figure 5.12 shows the structure of the new company. The group comprises a bank (*BNC*), two companies directly involved with the implementation of the Italian high speed railway programme (*Italferr Sistav* and *TAV – Treno Alta Velocità*), and two travel and service companies (*CIT* and *INT*).

The Italian railway company owns 373 million m^2 of land, of which about 2%

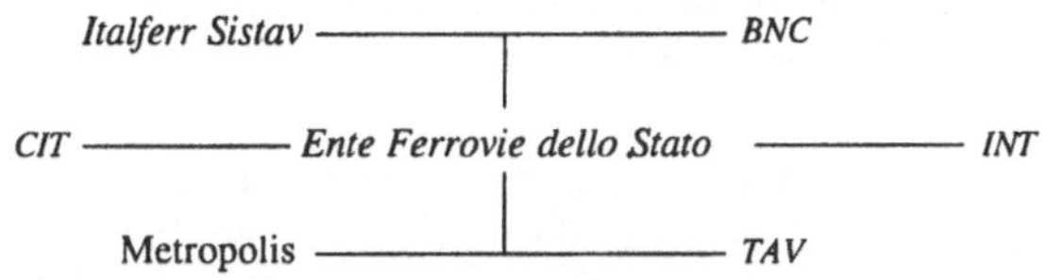

Figure 5.12 The Italian railway company group in the 1990s (*Source*: *Ferrovie dello Stato*, December 1991).

(7.45 million m^2) is accounted for by the surface area of the main railway stations, 17% (62.15 million m^2) by the land of the small and medium-size stations, and about 4% (16.0 million m^2) is vacant land mainly used for industrial purposes in the past (manufacturing and repairing carriages, diesel engines, etc.). The rest is occupied by the railway lines and by the service areas still in use.

In 1992 the *FS* owned a stock of buildings with a total gross floor area of 13.7 million m^2 and with a total volume of 68.1 million m^3. By 2001, according to the development programme, the property should increase to 21.5 million m^2 of gross floor area, and to 107.8 million m^3 volume. The bulk of the planned development should take place on the land areas of the main stations, which are also the first places to be served by the planned high-speed rail network.

The buildings of the main stations had a percentage weight of 17% of the total property in 1992, and this share is planned to increase to 30% by the year 2001. The *FS* has made an estimate of the market value of its property assets, of between 16460 and 21050 billion lire in 1992. According to the development programme, the property assets are expected to reach a value in the range 54200–62100 billion lire by the year 2001. It is expected that 36% of the planned value should come from the properties newly built or redeveloped in the main station areas.

For the whole period 1992–2001, the *FS* group has planned to invest 9220 billion lire for the restructuring of its real estate assets. Primarily, the investments are planned for the area of Milan (1850 billion lire), Naples (1150 billion lire), Verona (1050 billion lire), Rome (1000 billion lire). The direct investments on property assets in Turin exceed any other urban centre, as in this city the *FS* and the Commune of Turin have signed a development agreement, together with other local institutions, to implement a large-scale strategy for the re-use of the *FS* real estate assets in connection with the new city master plan. The strategy is centred on a swap between development rights and a vast area owned by the *FS* in a central location adjacent to the present headquarters of the Polytechnic of Turin. This area had been previously utilized for maintenance and repairs of trains, and in the city master plan it is assigned to the Polytechnic for the construction of new facilities. The *FS* has been assigned the right to build a mixed-use project (including one hotel, one shopping centre, offices and housing) on a plot next to the central station *Porta Nuova*. In addition, there is the major infrastructure project, whose cost exceeds 1500 billion lire, for Turin's new rail link. It consists of 15 km of rail lines placed underground in newly built tunnels, with six underground stations, including what will be the first Italian railway station for the high-speed trains. The first two phases of the project have been completed on time and on budget in early 1996. The third and last phase is scheduled for the period mid-1996 to end 2000.

Out of the total of 3600 billion lire of net revenues in the year 1990, the *FS* group obtained only 2.5% from its real estate assets, while the balance (97.5%) was generated by the train activities. By the end of 1999, according to the development plan of 1991, the real estate assets should count for no less than 55% of the expected revenues, estimated at 18300 billion lire.

In the next two decades, it may well be that the real estate development of many Italian cities will be led or greatly influenced by projects implemented on land owned by the railway company. This is a market trend comparable to what has been happening in other major European countries since the 1980s. Indeed, it appears that the presence or absence of a direct link with the European high-speed rail network will become a major factor in determining the ranking of European cities in the next few decades. These are also the conclusions of a study, *European regional prospects*, published by the British consortium Ereco in 1992. The train is again likely to play a major role in shaping the European cities, as it did in the course of the nineteenth century when the railway station became a landmark of European cities and a symbol of the first industrial revolution. In France, a study commissioned by the firm Auguste-Thouard (*Du bon usage du TGV*) in 1992 maintained that the *TGV* plays a positive role for existing business firms, although it is not sufficient per se to create new firms or new office districts. However, there is evidence that the *TGV* has greatly sustained the growth of an existing positive trend in the local real estate market. Apart from the well known case of Lille, where all the recent development has been centred on the new high speed rail station, in other cities (Tours, Le Mans and Lyons) the real estate markets were sustained by the link with the *TGV* network. In Belgium, the renovation activities around the central station, which took place in the first half of the 1990s, were also motivated by the prospects of the link with the French *TGV* system.

In Italy, according to the programme approved by the *FS* in December 1991, the diversification programme for the period 1992–2001 is designed in three phases. In the first one, the *FS* plans in each of the 14 cities mentioned above to set up a subsidiary company, through the controlled *Metropolis*, whose mission is to increase the market value of the real estate assets and to seek the participation of venture capital from local banks, firms and institutions. In the second phase, new projects are expected to be launched on the land owned by the *FS* group in each city. In the third phase, the development projects should be completed and the local rail link (*Pendolino* trains), with the Italian high speed network should be constructed.

As of 1996, the first phase has been completed in most of the 14 cities, whereas the second phase appears to be initiated only in some areas. Overall, the large-scale programme of the *FS* needs to overcome two obstacles to be successful. The first one is the establishment of a sound cooperative attitude with local governments and the local economic community. The second barrier to be overcome is the conspicuously low confidence of foreign investors and end-users. The two issues are interconnected. It appears that a large-scale development programme such as the one drafted by the *FS* group cannot be implemented without sizeable support from both national and foreign private investors.

Indeed, the involvement of foreign investors is crucial both for the investing capital, which could be directly mobilized by them, and for their experience with modern financing techniques for large-scale public infrastructures. For example, a project finance approach seems appropriate in the case of the planned 53 km rail tunnel under the Alps to link Turin in Italy with Lyons in France. According to recent estimates of the *FS*, construction would take 9 years and a repayment period would last 25–30 years. In this case, as the Channel Tunnel has demonstrated, any estimate of construction costs is merely indicative. The only sure thing is that the costs will be huge and that, in the absence of private venture capital, the green light to the project is at risk of being postponed indefinitely. Although the apparent lack of confidence of the foreign financial markets seems to be a major obstacle, it may well be that this issue will be solved almost automatically when the *FS* presents its plans for the construction phase in the near future. Still pending are the locations of some high-speed rail stations within the affected cities, such as Novara, between Turin and Milan. The solution is likely to come through a bottom-up planning approach which will seek the involvement of the local administrations, the business community, the political parties, and associations of consumers and residents.

CHAPTER 6

Case studies

6.1 Industrial re-use: the case of Bicocca in Milan

Context of the development

Bicocca, built in Milan in 1908, is the name of one of the oldest industrial plants of the Pirelli Group. Photographs dated 1922 exhibit Bicocca as a lively and busy industrial area, with its internal streets bustling with lorries, blue-collar workers, smoky chimneys and a train leaving the factories full of Pirelli tyres and cables. The area extends to about 680000 m^2 and is located in the northeast outskirts of the city, at a distance of 8km from the city centre (see Figure 6.1).

The case of Bicocca in Milan is a good example of a phenomenon that started in the early 1980s and which affects all the large Italian cities: the recycling of inactive urban industrial sites. Major industrial plants have ceased their historical function and released land ready for new uses. Pirelli production was shifted from Bicocca to other more technologically advanced factories, both in Italy and abroad; Alfa Romeo left Portello and consolidated its production in Arese. When in 1980–81 FIAT drastically changed its production systems, introducing the robots in Mirafiori and opening new plants in the South of Italy and abroad, the multistorey plant of Lingotto, with its fascinating test-track for cars on the roof of the building, soon looked like an outstanding piece of industrial archaeology. The re-use of old industrial properties affected not only car-related firms. For example, in Turin, Genoa and Naples the large plants of FINSIDER (steel) became available for new developments. Oil refineries in Mestre (facing the lagoon of Venice), around Milan, Turin and Genoa also became obsolete. The Italian state railway company (*Ferrovie dello stato, FS*) is one of the largest owners of urban areas previously utilized for industrial uses and now awaiting new functions. There are also many areas owned directly by the state, for example large military grounds well located in urban areas, which sooner or later will be re-developed.

However, the case of Bicocca holds a symbolic meaning for the role of Pirelli industries in the history of Milan. The re-use of Bicocca marks the end of the twentieth century, which had opened with the great promise of a new industrial society, a new way of production and of living. The industrial city now gives way to the entrepreneurial city, with new promises of wealth and social progress.

The content of the plan

The project area is 680000 m^2, of which 165000 (24%) will be covered and 512000 (76%) left open. The ratio between floorspace (572500 m^2) and land area (680000 m^2) is 0.84. Bicocca is a mixed-use project with the following components: 23% (132000m^2) for private housing, 40% (229000m^2) research and development; 32% (180500m^2) offices, and 5% (31000m^2) service functions. The main use is assured by the presence of the University of Milan, which took up the best share of the floorspace for research and development and for office use: 122000m^2 of floorspace, or 53% of the share allocated for research and development, with further opportunities for expansion. It should be noted that a share of the private housing must be *edilizia convenzionata*, that is housing to be sold with certain price caps in order to benefit from limited public subsidies.

According to the plan, the open space will be utilized as follows: 291000m^2 (57%) green areas, 77000m^2 (15%) sporting facilities, 46000m^2 (9%) squares and porticos, 98000m^2 (19%) roads. These figures are determined by the master plan of Milan and with regard to the specific requirements set for this area, which was declared a "special zone", as explained below.

In order to pay for the tax due to the city when a planning permit is granted (*oneri di urbanizzazione*), the promoter obtained from the city council the consent to provide, instead of the cash payment due, several buildings for public use (congress centre, libraries and community centres, multifunctional spaces) for a total of about 10000m per m^2 of floorspace (see Figure 6.2).

Payment in kind (infrastructure and buildings) rather than in cash has become almost standard practice for large-scale projects, because it brings benefits to both the private developer and the local authority. The developer acquires economies of scale in assuring the direct construction of facilities for public use, which in the end cost less than the taxes otherwise due to the commune. In addition, the developer can also control (in agreement with the public authority) the timing of the construction of these facilities, whose presence turns out to be a complementary ingredient of success of the development project itself. The commune, in turn, obtains the certitude that the project will be served by the level of public facilities prescribed by the law, without the hassle of a great deal of paperwork. The commune also avoids the risk that the tax amount received in connection with the granting of planning permits will be insufficient to build the facilities for which the resources are collected, because of delays in the utilization of the funds or because of unpredictable events (e.g. social emergencies and natural disasters), which may force the commune to utilize the funds elsewhere.

The owner of the property, the Pirelli group, promotes the Bicocca project through a controlled company, Milano Centrale Immobiliare, which is in charge of both project management and marketing. The selling price is, on average, 4 million lire per m^2 of floorspace. The global value of the project is estimated at 2300 billion lire. According to the available data, the re-use of the property does not entail special clean-up costs, as the soil was not contaminated by the previous industrial activities.

Accessibility to the project area turned out to be a crucial factor in the feasibility of the development; therefore, it will be analyzed in depth. Access to Bicocca via private transport is at present rather good. However, the expected increased volume of traffic has induced the public authorities to plan new roads. The major accomplishment is the so-called *Peduncolo*, a linkage between the highway ring

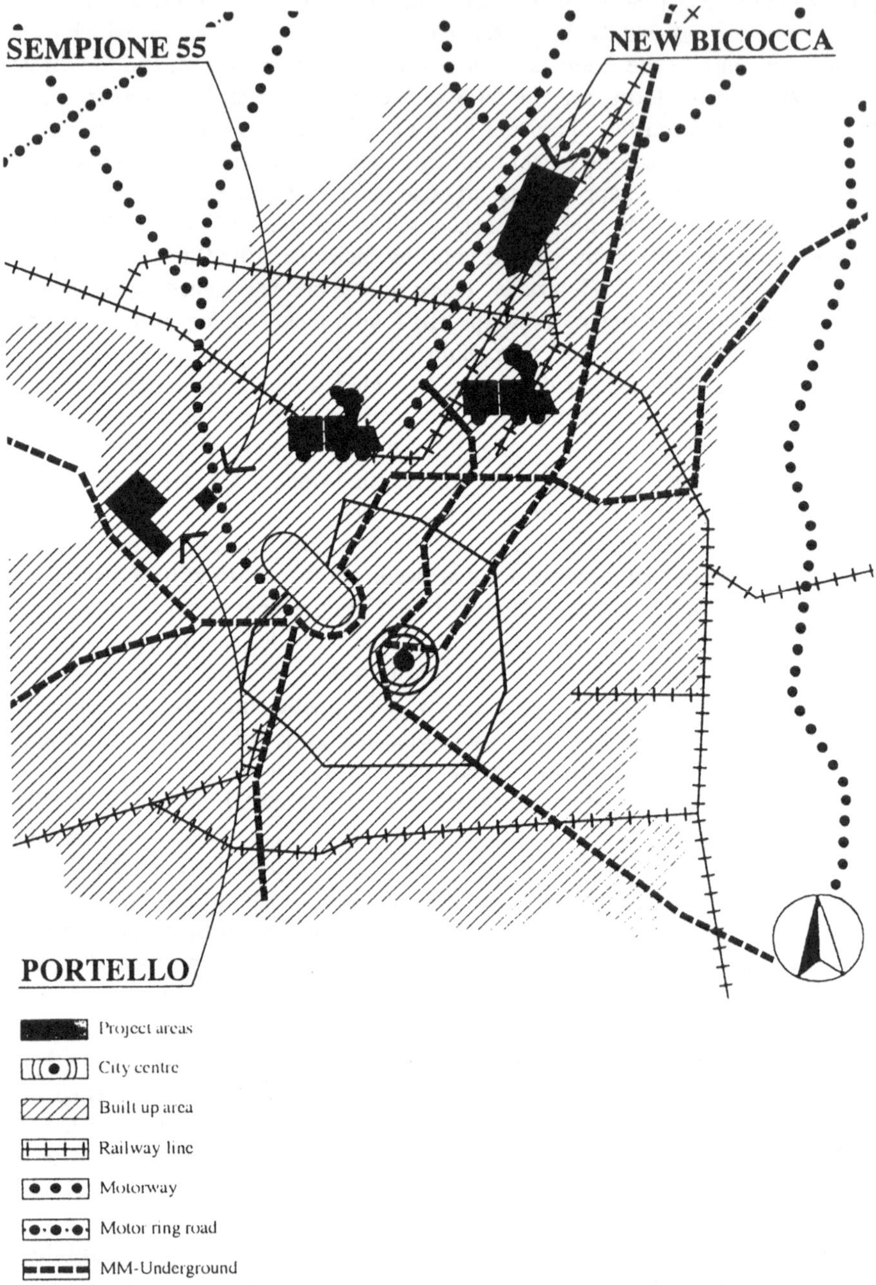

Figure 6.1 Location of three case studies in Milan.

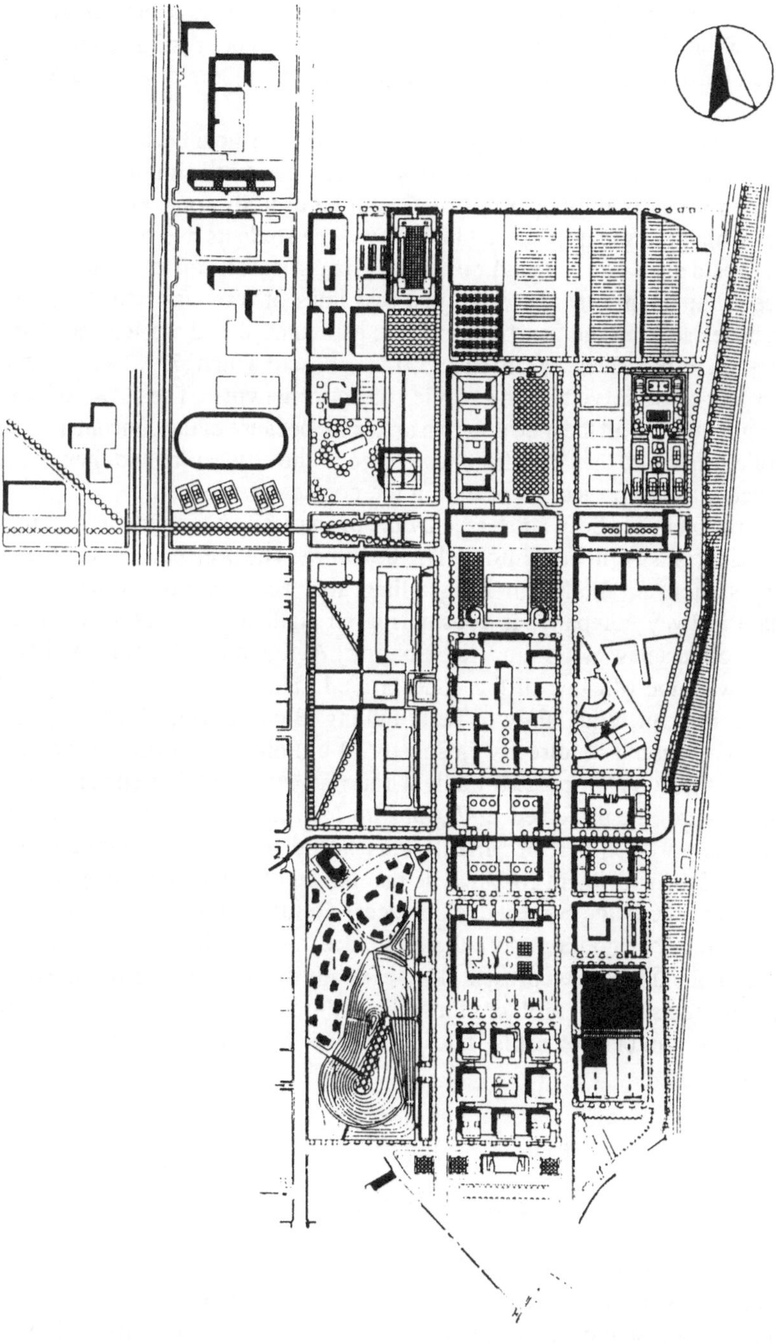

Figure 6.2 Layout of the new Bicocca (*Source*: *Milano Centrale Immobiliare Gruppo Pirelli*).

and Viale Fulvio Testi adjacent to the project site. It eases access to the Turin–Venice highway and the airport of Linate. The Commune of Milan plans to double some of the roads that give access to the area (e.g. via Sesto San Giovanni and via Cozzi), it also plans a new boulevard between viale Testi and viale Monza, and the enlargement of the bridge between via Porto Corsini and via Sesto San Giovanni.

The situation with respect to public transport is critical because of the anticipated load created by the new activities in Bicocca. According to the promoter, when the project is fully operational, it will be used by some 60000 people per day, of which the greatest share (44.5%) will be students and only a small share (6.5%) residents. The demand on existing mass transit infrastructures is likely to exceed the supply, as there are just two lines of tramway and three lines of urban buses serving the nearby area. It is estimated that there will be a peak of 20000 transits between 8.00 am and 9.00 am, most of which (73%) will be students who expect to be served by the public transport network. The flows of students will originate from all the metropolitan area and, because of the concentric network of public transport, they will first have to go to the city centre and then commute to Bicocca, as the project is located in the northeastern outskirts.

This problem has been addressed by the promoter, who maintains that the supply of mass transit will be reinforced in the future. The project is next to the Greco railway station, on the Milan–Monza line, and, according to the promoter, the FS (state railway company) is considering strengthening the downtown service and may open a new station in Viale Sondrio. At that point the railway line could connect with the M3 line of the underground. There is also the proposal for a light-rail line on Viale Fulvio Testi linking Bicocca with Garibaldi railway station. This would be directly financed (30 billion lire) by the promoter. Furthermore, there is a proposal for a light-rail line between Niguarda and Precotto, which is part of the projected *Collegamento Interquartiere Nord* (a semi-circular line for the northern part of Milan). This would link Bicocca with the M1 and M3 lines of the subway of Milan. The magnitude of the public transport projects confirms indirectly the current difficult accessibility issues. Some of the proposed infrastructures are nothing more than pipe-dreams, which may never come true. One may conclude that, at present, the accessibility of Bicocca is not satisfactory. Indeed, it seems more a problem for those end-owners, like the university, which attract large numbers of users, but it may well also be a serious hindrance for corporate firms.

Plan implementation

At the end of 1995 one third of the project (exactly 28.7%) had been built. The detailed stages of the implementation process are listed below.

The starting date of the implementation of the project was April 1985 when the Pirelli company, the promoter, signed an agreement with the Commune of

Milan, the province of Milan, the region of Lombardy, and the trade unions (*CGIL*, *CISL*, *UIL*), to transform the area into an "integrated and multifunctional technology centre". This was itself the outcome of a process that originated in the 1970s, when Bicocca became obsolete as a production plant. In September 1985 the promoter invited 20 international design firms to produce ideas for the re-use of the area. In the first phase the firms Gabetti and Isola, Valle and Gregotti were retained. Between July 1987 and May 1988 the Commune of Milan approved a change (*variante*) to the city master plan indicating Bicocca as Special Zone Z4, that is, an area where development was to be ruled by specific norms. In February 1989 the region of Lombardy approved the decision of the Commune of Milan to designate Bicocca as a special zone. In the second phase of the design contest, in July 1988, the project of Gregotti was chosen, thus becoming the base of future adaptations.

In March 1990 the Commune of Milan adopted the *Piano di lottizzazione* 1L and approved it in July 1992. This decision was ratified by the supervising public body *CORECO* in October 1992. In July 1989 the commune approved the *PIO* (*Piano di inquadramento operativo*), a plan to assure a link between the plan of the special zone Z4 Bicocca and the city master plan. In 1992 the plan for Bicocca was inserted into the 4° *PPA* of the Commune of Milan. In the same year the commune approved the request that the promoter should provide 20000 m^2 of land outside the Z4 at no charge to the city for public use. In May 1993 the *Commissione Edilizia* of the Commune of Milan made some observations on the aesthetics of the proposed design, but this resulted in only minor changes to the project. In September 1993 came the key decision of the city council to name Bicocca as the second university pole of Milan. Between December 1993 and February 1994 the request, submitted by the promoter, for a new *Piano di lottizzazione*, was approved. The new plan is named "1T". Between 1994 and 1995 the demolition works were completed and construction work began.

Thus, the implementation of a large-scale project such as Bicocca has required, so far, a ten-year-long series of public permits and plans. In the meantime, obviously, the property market has gone through various phases and when the promoter was ready to begin the actual construction works, in 1995, market demand was a different animal from that which existed when the implementation process began. The promoter, however, managed to obtain a high pre-let ratio (85%) for the commercial floorspace. By the end of 1995 this space had already been sold to investors. The percentage of 85% refers to the space allocated for research and development (which is 40% of the total) and offices (32% of the total); it does not include private housing, which account for 23% of total floorspace and whose construction has not begun yet.

The main investors are the following: banks, financial companies and pension funds (*ABF Leasing*, *BNL Leasing*, *Deutsche Bank Leasing*, *Fondo Previdenza CREDIT*, *INAIL*, *INPDAP*), public institutions (Commune of Milan, *AEM*, University of Milan), private firms and associations (*Associazione Cotoniera*, Digital, *Schiapparelli*, Pirelli). The list of tenants for the best part coincides with the investors but it also includes the following main end-users: research and development institutions (CNR, *Bic Lombardia*, *Incubatore Tecnologico Bicocca*, *Pirelli-Ricerca*, *Siemens*), press (*Rusconi Editore*, Reuters), public institutions (APT, Enea), banks and associations (*Cariplo*, *COMIT*, *Confartigianato*).

Conclusions

One cannot avoid a comparison between Bicocca and the well known case of London Docklands. The single main reason for the latter's failure, in the early 1990s, was poor accessibility via public transport. The project dramatically changed the carrying capacity required from the existing transport network. The new light-rail line, which was the only link between the project and the London Underground, was totally inadequate. Only when it was decided to extend the Orange Line of the underground to the Docklands did the market demand for properties in the Docklands area take off. Obviously, such an extension of the Underground involved huge costs, which were mainly covered by public financing, whereas the main benefits (in terms of added market demand) have accrued primarily to the private owners of properties in the London Docklands area.

Indeed, there are windfalls each time there is an injection of public money into new infrastructures as a direct support for a private real estate project. With regard to the classic works of urban and regional economics (e.g. Von Thünen, Christaller, Alonso), it seems clear that the true source of difference between rents of comparable projects built on different plots is *not* the location factor but the accessibility factor. The latter varies between plots, parts of the same city and also between cities and regions. It is well known that real-estate agents consider "location" the main factor (the three "L" factor) for success of a property, but if one looks carefully, it is evident that they use the noun "location" as a synonym for "accessibility". Location is a natural characteristic that cannot be changed. If this were the single success factor, the real estate market would have been divided once and for all into lucky owners and hopeless ones. Instead, accessibility can be improved by the action of public authorities through urban planning and the use of public money. Accessibility can also be decreased as the outcome, again, of a given public policy. It is as simple as this: what we could define as the three "A" factors (accessibility, accessibility, accessibility) determines the fortunes of almost all property projects. Through urban planning, the public authorities determine the levels of accessibility of all urban areas and ultimately determine the making of urban land values and real estate fortunes.

A case in point is the town of Lille, which has a poor location, being located in a corner in the northeast of France, in a region with no special characteristics. Neither the private owners nor the public authorities could change this geographical fact. But they were able to change the accessibility through public mass transport to the area, and this is the most important single factor that reversed the sluggish real-estate market in the area, making it a suitable alternative to other markets, notably Paris.

Another important element in the Bicocca project is the presence of important public institutions as end-users of commercial spaces inside the project. The location of the University of Milan, CNR, AEM (electricity company of the Commune of Milan), ENEA (national institution for nuclear energy), and APT (office for tourism promotion), has made it possible for the promoter to maintain that the investment of taxpayers' money in improving the accessibility of the area is in the interests of the general public and not simply an integral factor in the market success of the project.

It seems appropriate to question how this project was really market orientated, in the sense that its implementation became possible only when the promoter was assured of a significant demand of space from public institutions, rather than private entrepreneurs. Public bodies are not known, until recently at least, to be good negotiators. In other words, the size of Bicocca and of similar projects is so large that, in a saturated market such as that of Milan, demand can come only from the public sector. However, it is often the case that public institutions do not behave like corporate firms and may value differently the pros and cons of property deals.

Conceived initially as a technology pole, Bicocca has shifted to a mixed-use project with a predominant use of the space assured by public institutions such as the University of Milan and CNR. This is a common feature of most so-called "technology poles", which were so fashionable in the early 1980s and which, when implemented, turned out to be almost always mixed-use projects. They captured little or no private research and development from the commercial firms that were expected to flock to them.

On the positive side, it must be noted that this case concerns a project that is actually being implemented. Together with Lingotto in Turin, which is almost complete, this is a rare situation for the large-scale projects of this nature, which were conceived in the Italian property markets of the 1980s and 1990s. The city of Milan is undoubtedly better off, now that the case of Bicocca is no longer a problem of the past but a promise for the future. Nevertheless, as always in these cases, one should ask what the real costs of the project are for the taxpayers, and whether the public resources invested, for example in improving the accessibility to the project site, could have been utilized more efficiently elsewhere in the city, or could be claimed in a greater share from the promoter.

The chances are that, because of the presence of key actors, in financial terms the project will prove a highly positive venture for the promoter. The University of Milan, which acted as anchor tenant, attained financial autonomy only in the 1990s but when the project was conceived the promoter could assume that the university was ultimately backed by the financial might of the state. By the same token, it is questionable whether Bicocca could be taken as a model for recycling other large industrial areas, for example, the Bagnoli plants (steel) in Naples, or Porto Marghera (oil refineries) in front of the lagoon of Venice. It is perhaps too early to have a comprehensive view of this case. Conceivably, the final verdict will be influenced by the way the new Bicocca functions and repays the city in terms of "image dividends".

6.2 The recycling of old industrial properties: the case of "Sempione 55" in Milan

Context of the development

Sempione 55 is the name of the real-estate project built within an industrial area located in the northeast of Milan, facing one of the major radial axes (Viale Sempione) of the city and leading directly to Castello Sforzesco, a landmark of Milan (see Figure 6.1). For decades, the area had been used as the main branch of *FIAT* Auto in the city. In 1982, as part of an overall restructuring of all Italian branches and subsidiaries, *FIAT Auto* decided to convert the property, since its industrial and commercial activities had been moved to other plants. This decision had matured in the late 1970s, and was also influenced by the growing contrast between the industrial activities traditionally conducted on the premises, and the characteristics of the prime location that the area had assumed over time. In addition, in the early 1980s, *FIAT Auto* embarked in a structural overturn of its core business method of production. This effort required the investment of unprecedented financial resources. At the same time, there were clear and unconditional signs that the real-estate market of Milan was entering a boom period for offices. All this led almost automatically to the decision to convert the industrial area of Sempione 55 into an office centre.

The irrational climate of the real-estate markets of the early 1980s may be difficult to recall now, since the slump of the 1990s is not yet over. But in Italy in those years most real estate agents believed that the market could go up and up for ever, and regarded expectations of price rises in the range of 20% per year and even higher to be business as usual. After all, they maintained, a property is real wealth, compared to the unpredictable worth of company equity assets and to the declining value of the Italian lira, which had experienced years with two-digit inflation rates. The housing market of Milan, as in the other major cities, was still growing on the unexpected effects of the national law which introduced fixed "fair rent" levels in 1978. Housing for rent disappeared from the market, whereas housing for sale benefited from an extra growth of demand generated by those who were forced to buy a "roof", as no other solution was available to meet their housing needs. The national economy was gradually shifting from an industrial base to a service economy. In this context, the service sector and the property it occupied appeared to guarantee the future wealth of cities.

The content of the plan

The property extends to about 23000 m^2 in a unique prestigious zone, a place of good-quality housing and businesses, with a likelihood of further commercial growth (Fig. 6.3). The city subway does not serve the neighbourhood at present. Nevertheless, the market appeal of the initiative was shown from the immediate response from prospective clients. The building, initially conceived for sale at completion of the construction works, found qualified buyers within a few months of the start of construction.

This is a distinctive factor in this case. In fact, the first idea of the promoter

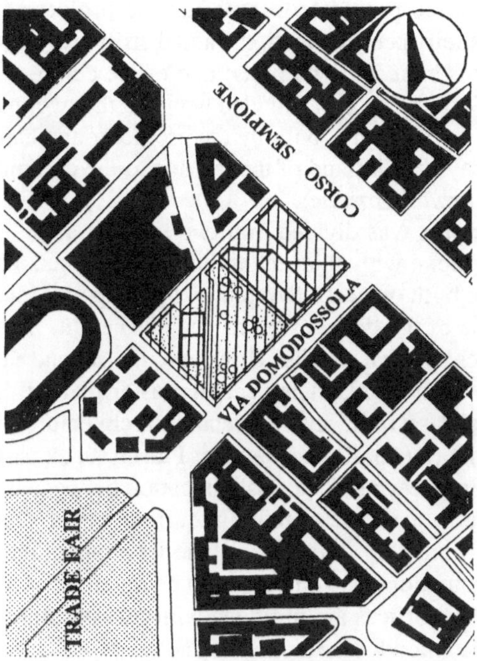

Figure 6.3 Layout of the *Sempione 55* project in Milan.

was to convert the property into small and medium-size offices. The initial planning permit and the detailed architectural designs all referred to this choice. Incidentally, this choice was in line with the prevailing attitude of real-estate firms in the early 1980s. The underlining idea was that the market suffered from an insufficient supply of office space, whereas the demand was thought to be an ever-growing alien beast. The truth was different, of course. The private developers learned the hard way (i.e. unsold and unlet properties) not to ignore the actual requirements of the market. It soon became clear, in Milan as elsewhere in the country, that there could have been a lack of supply, but only of quality office buildings in locations with excellent accessibility and urban infrastructures. It also became evident that the development of office centres for generic future clients was a risky business because office work has differentiated over time. In the 1990s the technical requirements of modern office work vary greatly from firm to firm, so an office building cannot accommodate efficiently any kind of business. In brief, there is no longer, if it ever existed, a "standard" business office centre that can be developed and successfully offered to the market.

In theory, the conversion of the property into many small and medium-size offices would have been highly profitable. However, the promoter received an expression of firm interest in the project from two prime potential clients: the Bank of Italy, which was looking for new headquarters for its branch in Milan, and IBM *Italia*, which had a similar problem. The agreement with these two prospective clients led the promoter to redesign the project completely, so that it would respond to their specific functional and distributive needs. In the final solution, the property is split into two parts: one office centre comprising two main buildings (the bulk of the project) divided between the two major clients, and one sports centre plus a small park, built by the promoter and handed over to the city of Milan as part payment in kind of the tax due to the commune for the planning permits (*oneri di urbanizzazione*).

The office complex was divided into two units to abide by height limits and to ensure a better blend with adjacent architectural forms. Thus, the first part of the office centre is built on corso Sempione, lined up with the existing buildings, and it retains some of the typical aspects such as height, materials and the repetition of a few architectural details. However, the second building is a conventional block of steel and glass. The general master plan of the project was done by engineer Lucio Passarelli, and the detailed architectural designs are those of architect Gino Valle (IBM part) and engineer Pierini (bank part). The general contractor was *FIAT* Engineering. The final allocation of space is given in Table 6.1.

Under pressure to meet the exacting requirements of such prominent clients as the Bank of Italy and IBM *Italia*, the developer applied the most advanced concepts of office automation. The outcome can be regarded as an example of "intelligent building". The offices are provided with floating floors and false ceilings, which facilitate the smooth distribution of cables, pipes and all the technical networks that are needed today. Most functions of the office complex are managed

Table 6.1 Detailed allocation of space for the *Sempione 55* office development in Milan (gross floorspace, m^2).

Total area	23000
Building volume	70000 m^3
Sports facilities (public)	
Sports centre and public park	
Surface area	11370
Sports area	
Area underground	250
Area above ground	2515
Office centre (private)	
Part occupied by IBM *Italia*:	
Gross surface area underground	12090
Gross surface area above ground	17930
Part occupied by the Bank of Italy:	
Gross surface area underground	3240
Gross surface area above ground	4100
Internal roads	1150
Car ramps	880
Internal squares	5180

Source: *Attività Immobiliari* (FIAT Group).

by two computers located in a control centre, from where the technicians can ensure the safety of the complex without leaving the room. The computerized control system minimizes energy consumption and adapts the energy requirements to the varying conditions of use (normal business hours, weekends, etc.) of the complex. For example, not only can heating and cooling be adjusted independently room by room, but also the quality of air is continuously monitored to prevent internal pollution generated by computers, other office equipment and smoking. Special light sensors adjust the intensity of lighting to the degree of natural light and other parameters.

The Bank of Italy occupies the first three floors above ground of the building on corso Sempione. The vault, the technical centres and the local deposits of the bank are sited underground. The IBM offices are also found in the three floors above ground in the building on corso Sempione and from the second to the seventh floor of the building behind, whereas the ground floor and the first floor of this second building include board rooms with a capacity for 12 to 20 people; an auditorium with 96 places to sit; a computing centre (about 450 m^2) with common services and building; a control room, reception and staff functions. On the eighth floor of the back-building there is a restaurant. Two floors underground are meant partly for parking (244 places) and partly for storage. Underground, moreover, there are the thermal, refrigeration and electrical centres.

Plan implementation

The area belonged to *FIAT Auto* until December 1982, when it was transferred to *Attività Immobiliari*, a company in the *FIAT* group that specializes in real estate promotion. Before the sale, the area had already been reclassified in the city master plan as *Zona TA7*, that is, an area suitable for office use instead of industrial use, under the obligation to sign up an agreement with the commune (*convenzione*) and to propose a plan of allocation of spaces for the property (*Piano di lottizzazione*, *PL*). Having abided to these obligations, in July 1983 Attività Immobiliari obtained from the commune the planning permit to build 70000 m^2 (gross) of offices.

It must be noted that, as part of the *PL*, the promoter had to give to the local authority about 11000 m^2 of land (out of the total of 23000 m^2). In addition, on the land set aside for public use, the promoter built a sports and cultural centre as a part payment of its dues towards the commune. As the land accruing to the local authority should have been more than 11000 m^2, the difference was converted to a payment of about 1 billion lire to the local authority. The project also entailed the tax due for planning permits (*oneri di urbanizzazione*), which is proportional to the gross floorspace and to the volume of the commercial properties whose construction is authorized. This amounted to about 3 billion lire and, as is fairly common in these cases, the promoter was granted the permission to build public works for an equivalent amount.

In November 1983 the commune granted the permit to demolish the existing buildings. Demolition work was completed by February 1984, whereas the construction activities of the office centre began in May 1984. The promoter sold about 20% of the floorspace to Bank of Italy in July 1984. The contract with IBM *Italia* was signed in December 1984, covering the remaining 80% of the office complex. Initially it was a leasing contract; then, in July 1987, it was substituted with a full sale contract. Subsequently, IBM sold its share of the office centre to *CartaSi*, the leading Italian credit card company. The office centre was completed in April 1987. The developer obtained the planning permit for the adjacent sports and cultural centre in July 1984. The works started in January 1985 and were complete in June 1986. The planning permit for the public park was issued February 1986 and the works took until July 1987, when the full project could be considered complete.

Conclusions

The conversion of the industrial property of Sempione 55 into an office complex and leisure facilities was a commercial success. It seems that the implementation of the project left both the promoter and the community around the project site better off. This is possibly an ideal goal for all serious real estate firms. Among the success factors, the ability of the promoter to act with the right timing was crucial. The company *Attività Immobiliari* was able to get in and out the office market very quickly, so that when the office market of Milan began to decline at the end of the 1980s, the sale of the property had already been completed.

The time factor is usually underestimated in property development, whereas it is one of the most influential components of the financial viability of any project. Developments completed rapidly are the safest: the promoter does not then suffer the unpredictable reversals of the market and also avoids unexpected changes in the planning schemes, with consequent delays in obtaining planning permits and financial credit from banks. The promoter initially acted in a fairly conventional way for those years, trying to convert the property into something (small and medium-size offices) of which the market was apparently in need. Yet, in the early stages of the project, the promoter was flexible enough to change its existing designs to adapt the project to the needs of clients (Bank of Italy and IBM *Italia*) unforeseen when the first idea to convert the property was initially conceived.

From the viewpoint of the local authority, the project can also be considered to be a positive change in the urban landscape. It enhanced the image of the area, brought to the public domain an area of 11000 m^2 with sports and cultural facilities, and also provided the commune with a sizable amount of *oneri di urbanizzazione*, the tax due on issue of planning permits. With the approval of *ICI*, the local tax on real estate properties, in 1993, the project became an added tax base for the benefit of the municipal budget.

One may also see in this example, however, a good case to question the current concept and use of the revenues due to the local authority whenever a planning permit is granted. Currently, the law requires the collection of revenues for "building" things (parks, roads, sewage, primary schools, etc.) without taking into account the cost of maintaining them to a set standard for a given number of years (e.g. the expected life-span of a given project). The use of these "betterment levies", may result in a direct benefit to the project of the promoter, and only indirectly a benefit for the community. In the case at hand, for example, the public park is obviously a benefit for the entire neighbourhood, but its presence offers perhaps the most sizable benefit to the office buildings facing it. Although the promoter carried the cost of developing the park, the municipality has the burden of paying for its maintenance (lighting, security, green planting and cleaning). Thus, it would be desirable to take into account (in computing the *oneri di urbanizzazione*) the future maintenance costs, in the interests of both the municipality and the promoter. In the event of a budget deficit, the commune would be forced to cut the regular maintenance expenses, and the market value of a property facing a rundown open area (which was once a pleasant park) would inevitably decline.

6.3 The new central business district of Naples

Context of the development

The initial idea of building a new "business district" in Naples goes back 1962, when the Commission for the new city master plan identified the area of *Poggioreale* as the most suitable for its strategic location in relation to the city centre and to the main communication infrastructures (see Figure 6.4). Throughout Italy, the reconstruction phase after the Second World War had been completed in the mid-1950s. Since then, the major efforts of the building industry were directed to the expansion of the housing stock. Swelling outskirts of the largest metropolitan areas housed thousands of new immigrants. It was not until the 1960s that the idea of having a new "business district" became a common feature of a new generation of city master plans. These "districts" had invariably some common features: a location adjacent to the historical centre of the city; the presumption that it would be an ideal location for most public activities; a lack of direction on what to do with the large areas that would have been vacated in the core of the urban fabric if the new business district had actually been built; and a disregard for the actual market demand for office space. It did not came as a surprise that the vast majority of the proposed business districts remained on paper.

In the case of Naples, the idea obtained official approval in 1972, when the proposal for a new district was incorporated into the general master plan (*PRG*) of the city. The project was seen as the creation of a new area within the city, through the redevelopment of an abandoned industrial zone. The aim was to bring new life to that zone, while relieving the pressure of housing demand, commercial activity and traffic on the historical city centre. From this viewpoint, the chosen location, at *Poggioreale* on the edge of the city centre and bounded by the railway lines entering the central station, was considered suitable. It also facilitated better connections with the outlying districts of Naples and the rest of the region.

The designated area is at the centre of the regional transport network, with direct links to the motorways and other transport systems. Promoters of the

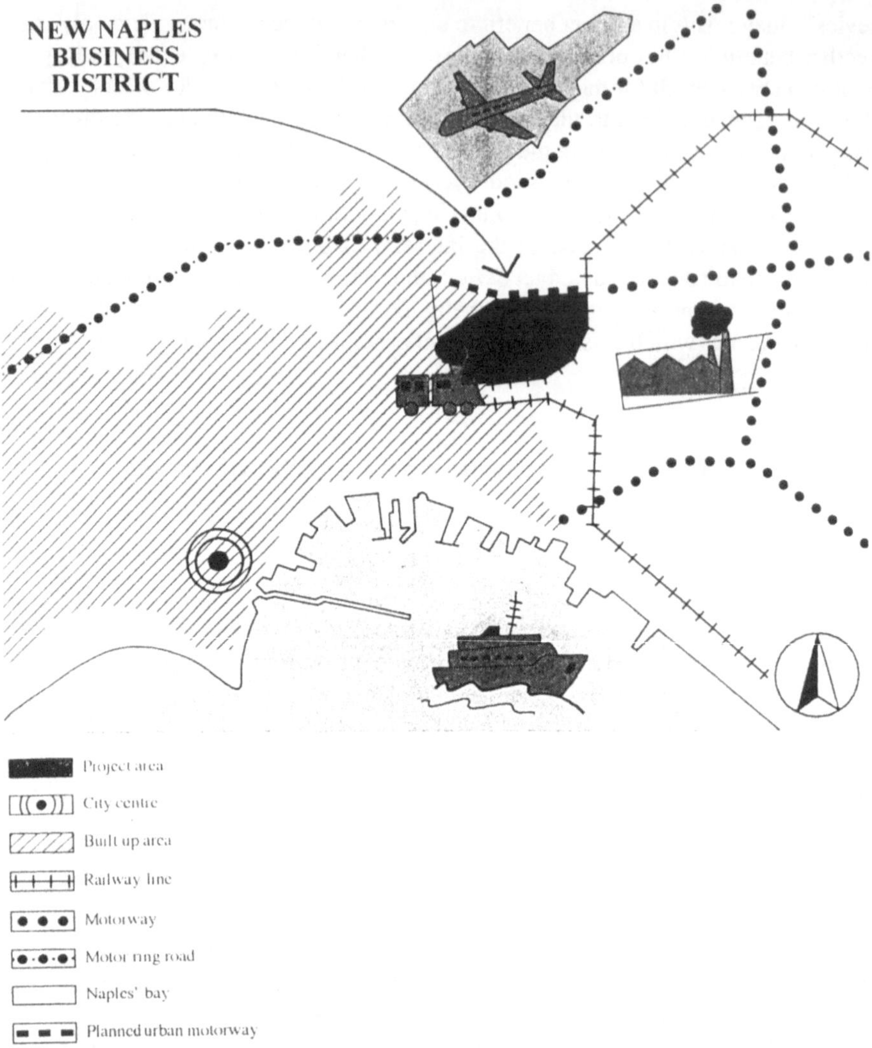

Figure 6.4 Location of the new business district project of Naples.

proposed business centre stressed the fact that it was not bound to become a quarter separate from the rest of the city, but rather it was an integral part of it. The main benefit the project was expected to bring would be the reduction in business overcrowding in the historical centre. In fact, the project called for the relocation of the headquarters of major public institutions, traditionally housed in the heart of the city. This was expected to bring a significant reduction of traffic flows, thus allowing the city's historical centre to breathe once more, and lay the basis for its subsequent restoration. Indeed, the creation of the business district was presented as a remarkable public project, one capable of restoring to the city its former role as the capital of the Italian South, and one of the greatest cities of Europe.

The content of the plan

The project for the new Naples business district covers a total area of 110ha, split as follows: on the western part 50ha owned by *ITALSTAT* through its controlled company *MEDEDIL*, and on the eastern part the remaining 60ha owned by public institutions, including, first of all, the Commune of Naples (Fig. 6.5). The area comprised several poorly maintained warehouses and commercial buildings, substandard housing and vacant land. The project foresaw a total building volume of 6220000m^3, of which 3250000 m^3 concern the area owned by *MEDEDIL*. Of the project building volume, 80% was to be allocated to office and retail use, whereas housing had a minor share (20%). The project is served by several public and private parking areas, providing a total of 25000 car spaces.

The plans for the new Naples business district were drawn up in 1982 by the architect Kenzo Tange, following some simple fundamental principles, which are in brief: the separation of pedestrian and motor traffic; public pedestrian areas organized along three longitudinal axes (the "green" axis, the "public" axis and the "sports" axis) with transverse linking structures containing rail transport and other facilities; the adoption of a high building-density index with high-rise residential buildings (50m, 70m and 100m) accounting for 30% of the total building volume (this choice was proposed as an exception to the height limits imposed by the *PRG* and caused several debates in the city council); and a strong emphasis on the provision of green spaces, which in practice occupy half of the total site area. They now create an environmental quality unusual in modern urban developments.

The "green" axis, consists of vast "hanging gardens" measuring 70m by 800m. It was designed by the architect Pier Luigi Spadolini, who has provided extensive rest and recreational features. Like the other axes, it is closed to vehicle traffic, which circulates at a lower level. The "public" axis, as long as the "green" axis, has two large squares at each end, surrounded by key public offices, the headquarters of Campania region and the law courts. The "sports" axis is parallel to the other two axes and is equipped with an Olympic-size swimming pool, tennis courts and recreational facilities. It lies between residential blocks and next to a park with an artificial lake.

Vehicle traffic flows along the lower level road network, which gives access to the central buildings and to the public and private car parks. From the private car parks there is access to the buildings; from the public car parks, escalators and lifts lead to the pedestrian levels where the entrances to the buildings are located beneath porticoes lined with shops, businesses, bars and restaurants. Access to the area is via two road networks: the main high-speed system directly connected to the regional and national motorway networks, and a secondary system penetrating the complex, giving access to individual buildings and areas. There are two railway stations within the complex, providing connections with the greater urban area of Naples. In conclusion, the plan of the Naples business district aimed to be an example of a multifunctional urban development that could combine service sector (office) functions with residential buildings in a complex with excellent services and a pleasant urban environment.

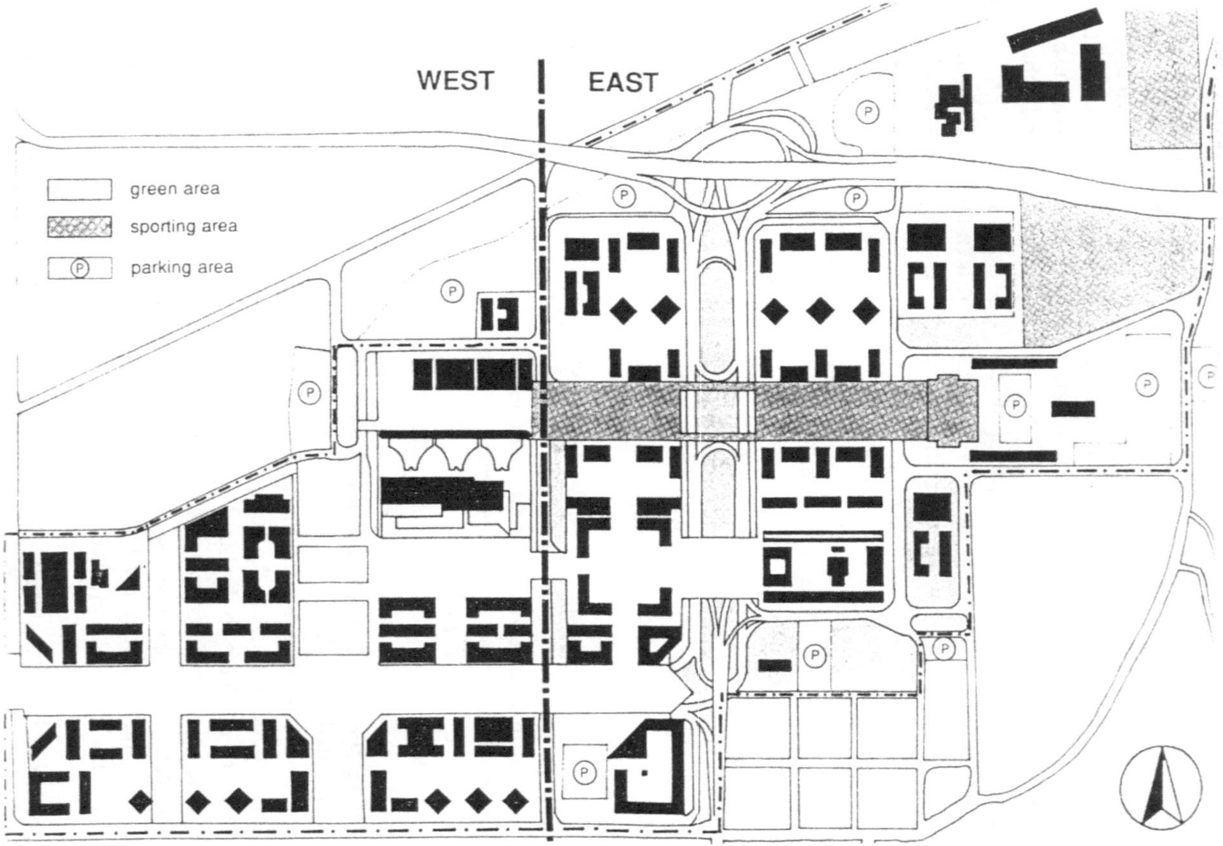

Figure 6.5 Layout of the new business district project of Naples.

Plan implementation

The project was promoted, implemented and coordinated by *MEDEDIL - Società Edilizia Mediterranea*, a construction company controlled (80%) by *ITALSTAT* (later transformed in *IRITECNA*), the major public company specializing in large infrastructural works, which is owned by the giant public holding *IRI*. Between February and October 1982, *ITALSTAT* acquired all of the shares in *MEDEDIL* (increasing the company's share capital from 2 to 50 billion lire) and took direct control of the firm. Under *ITALSTAT*'s management of *MEDEDIL*, the project entered in a phase of real implementation, after a decade in which it was discussed and modified fruitlessly several times.

In 1982, *MEDEDIL* carried out a major revision of the existing preliminary master plan for the new business district and engaged the services of the architect Kenzo Tange who completely revised the designs. This marked the start of the rapid implementation of the large-scale development plan, which was, at that time, one of the largest in Europe. The promoter set up a specific project management office to deal with the highly complex task of implementing the project. Between 1982 and 1985 the promoter completed the detailed drawings of the buildings and of the infrastructure networks. As the technical documents were ready step by step, the application to the Commune of Naples for planning permissions began in 1983. The permits for the construction of 56 buildings were granted in 1983 and onwards.

It must be noted that the action of the promoter concerned just its own area, that is the western part (50ha) of the global area (110ha). The initial project also comprised the 60ha plot on the eastern part, owned by local public authorities. The construction works began on the western part in 1983 and were completed in 1991, whereas the eastern section of the project (initially scheduled for completion in 1996) did not enter into a real implementation process. At present, there is little expectation that the eastern part of the project will ever be initiated. Nevertheless, according to the promoter, the project completed by *MEDEDIL* on the western part was a commercial success, as most of the offices were completely sold before the completion of the construction phase, and the 750 housing units, built after the offices, were also a commercial success. *MEDEDIL* has ensured since the opening of the district an internal security service and a private service for the collection and disposal of garbage, plus a maintenance service of a high standard for the public amenities (e.g. green areas, fountains, street lighting) within the boundaries of the project.

To develop the new Naples business district, the promoter obtained financing from various sources. The single most important financial backing was a loan of 110 billion lire at a favourable interest rate from the European Investment Bank. The application process took between 1984 and 1986. Repayment of financing covers the period 1989–96. The total cost of infrastructure works was about 300 billion lire. Concerning the sale prices, the cost of serviced lots was in the range of 180–360 thousand lire per m^3 of construction. The cost of the completed buildings was on average 600000 lire per m^3. Thus, the weight of the serviced land price can be estimated in the range between 25% and 60% of the price paid by the end-users.

The construction of the new business district directly created 1500 jobs. The project provided indirect employment for a further 800 people, creating a total of 2300 jobs in all. It is estimated that the completion of the project on the eastern part of the area, including all infrastructural and building work, would provide, if ever carried out, 3000 jobs per year for five years.

The promoter managed to present the new business district as a convenient area for the relocation of the headquarters of key public institutions, such as the region of Campania and the Law Courts. This was probably the single most important marketing action conducted by the promoter (*MEDEDIL*) and its holding company. In fact, these prime institutions attracted many other public bodies and private offices, which could benefit from being located next to them. In addition, the presence of such prominent public institutions became a valid argument in negotiating an agreement with the Commune of Naples concerning the underground extension of the *Circumvesuviana* railway line beneath the business district. This improved the accessibility to the project from the whole of the metropolitan area and consequently it increased the market value of the project buildings.

In conclusion, the Naples business district has been a reality since the early 1990s and is regarded as a lively new part of the city by the mid-1990s. Apart from the region of Campania and the Law Courts, the client list of the business district comprises most of the key companies of Naples, including: *Infrasud Progetti*, *SME Finanziaria*, Olivetti, *IMI*, *Isveimer*, *Banco di Napoli*, *Amministrazione Postale*, *Azienda di Stato per i Servizi Telefonici*, *Istituto Universitario Navale*, *Consiglio Nazionale delle Ricerche* (*CNR*), and *ENEL*. Some private professional studios and offices have opened in the district, and there are also many residents.

Conclusions

It seems remarkable how this project is little known not only abroad but also in Italy. Yet, it is one of the few large-scale projects in Italy that was not only designed but actually built in the past 20 years. Evaluations of this project are mixed. It is certainly a success for the developer, when taking into account the challenge to deal with a development proposal that had been around for many years with no practical results. This project has been seen (especially by many citizens of Naples who work abroad) as a reason for pride in the city. By the same token, the project cannot be considered a complete success, as only 50% of the huge available area (the western part, owned by the developer) has been developed (for example, the region of Campania did not locate all its headquarters in the new district). Even so, the volume of the built project exceeds 3 million m^3 and one may wonder whether the demand of the real estate market of Naples would have been capable of sustaining a further 3 million m^3 of office space and related commercial activities. It may well be that this shortcoming of development is a blessing for the city, which has retained a vast area where it can, for example, locate future strategic infrastructures.

The project has not been praised for the quality of the architecture, which has been seen as completely foreign to the urban fabric of the city. Indeed, the employment of world-famous architects can be regarded as a rather provincial attitude, when it appears to be a means for overcoming local resistance to a development proposal. The very scale of the project is bound to make the development a landmark for the city for many decades ahead. This fact should have suggested, perhaps, the launch of a contest for attracting the best architectural ideas from national and international architectural firms.

From the public policy viewpoint, this project seems to have absorbed most of the attention of the Commune of Naples in the urban planning field, whereas other parts of the city, notably the historical centre, continued in a slow decay. In fact, the transferral of office functions from the city core to the new district is not enough to give way to a virtuous circle in the refurbishment of the historical areas. From this viewpoint, Naples is a perfect example of the fact that the problems of a large city cannot be tackled on a project by project basis, but rather require a comprehensive view and a general planning scheme designed to respond both to market trends and social needs. The downturn of the city was halted only with the positive efforts of the new city administration elected in 1993 and headed by the Mayor Bassolino. The new administration proposed a sort of renaissance programme for the entire city, including new functions for the old industrial area of Bagnoli, and the rediscovery of the historic role of a "European culture capital" for the city of Naples.

The final judgement on the new Naples business district will only be possible in a few years time, when the scheme has been functioning for a while, thus enabling an evaluation to be made in terms of its efficiency as a business centre. Another key trial for the new district will be to keep up to a good standard the maintenance of its buildings, accessibility and services. It will also be interesting to see if the clustering of new activities from the core city in the new district will generate new real estate developments and commercial enterprises in the surrounding areas. This will be a decisive issue, if the new project is to blend with the city and act as an innovative event for the urban wealth of the whole city, instead of becoming an alien part of a once-united city. Finally, as always in these cases, the true judge will be the market, whose response so far has been encouraging.

6.4 Industrial re-use: the Portello project in Milan

Context of the development

Portello is the name of the renowned Alfa Romeo works in Milan. It became the first case of large-scale redevelopment of a disused industrial site in the city. The history of Alfa Romeo is interwoven with that of Milan, and the analysis of this case requires one step backwards to place the case in the town-planning context and in the correct historical perspective. Between the late nineteenth century and the first years of the twentieth century, Italy entered the restricted league of industrialized countries, amid economic and social contrasts that have characterized its dual growth. The motor sector was the fastest rising industry in a national context where the industrial production output doubled in the span of 15 years. Two companies assumed a prominent role: FIAT, founded in Turin, and Alfa Romeo, located in Milan.

Via del Portello was a street in the northwestern outskirts of Milan when the Italian car company Darracq built a large works in 1906. In 1910 the property became the first large industrial plant of the *Anonima Lombarda Fabbrica Automobili* (ALFA). The factory was technologically advanced and had a gross floorspace of 7000 m^2 and 300 highly specialized workers. The company was then bought by Romeo during the First World War, changing the name to Alfa Romeo. As the production expanded to aviation engines, trucks and even bullets, compressors and perforators, the area occupied swelled to 160000 m^2 at the end of the First World War. In the post-war period, the company was saved from bankruptcy by state intervention in 1926 and finally it became a satellite company of IRI, the public-interest holding company that was founded in 1933 in a major reshaping of the Italian banking system after the 1929 collapse. In the second post-war period, the Portello area had reached its present size of 230000 m^2 and could no longer be expanded, because the city had also undergone an unprecedented enlargement and had surrounded the factory. It was in 1959 that Alfa Romeo bought a 2 million m^2 area in Arese, a small town near Milan, to which all car production lines were transferred. At that point the re-use of the old property of Portello became an issue of recurrent debate in the city. As the decay of the vacant buildings inside the Portello area accelerated over time, the economic base of Milan changed through the 1960s and 1970s from industry to commerce, trade, finance and services. This led the property to be seen as an ideal place for the expansion of the adjacent Milan Trade Fair (*Fiera di Milano*).

The Milan Trade Fair, whose origins date back to the first International Exhibition of 1881, was given a definitive location in 1922 when the municipality chose a vacant plot of 630 m by 530 m, next to the Alfa Romeo factory of Portello. This physical co-existence, side-by-side, of pure industrial production and commercial promotion and exchanges became a typical feature of the economic base of the city up to the present day. The relationship between the Milan Trade Fair and the city has been for some time a complex problem and one that has not yet found a satisfactory solution. Indeed, the Fair is situated within the city itself. It is one of the few in the world to form such an intimate part of the urban fabric. This fact has posed serious problems both to the city and to the Fair. The city had to tackle the urban traffic problems generated by the Fair's visitors (over 5 million per year in the early 1990s), and the Fair's need for new space has been growing. But the surrounding city had closed off any opportunity for expansion, with the exception of the Portello area (Fig. 6.1).

In Milan, the economic development that took place in the 1980s produced radical changes in the economic and social life of the city, revealing the inadequacy of the physical context in which modernization and growth was taking place. Idle industrial areas left grey gaps in the texture of a city that increasingly needed services and facilities to restore balance and harmony to the life of its

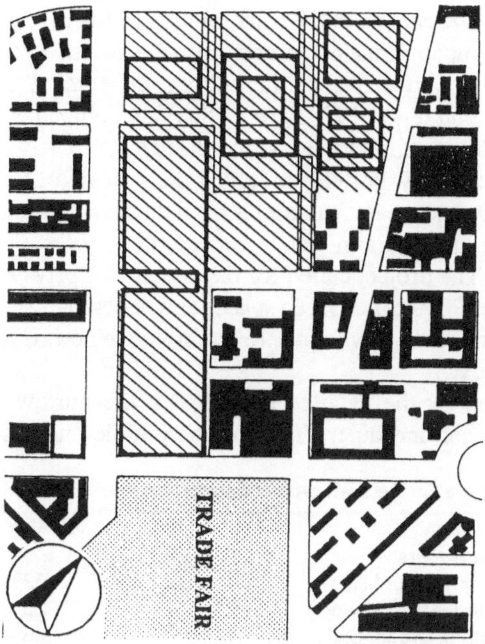

Figure 6.6 Layout of the Portello project in Milan.

inhabitants. The sudden onslaught of the automobile culture left the city in a confused state between a past that was rapidly disappearing and a future that was unknown. Some industrial cities had reacted by promoting a process of modernization, whereas others had sought to recreate the links with the past in order to grow within an historical framework. Milan seemed to have chosen the second course, through the re-use of fragments of its past, scattered throughout the city. The Portello project appeared as a golden chance to define a piece of the future growth of the city. To this end, the city council of Milan approved a change to the existing city master plan in 1985, followed by the approval of the detailed planning scheme, *Portello Sud-Fiera*, in July 1989.

The content of the plan

The project entirely covered the land formerly occupied by the Alfa Romeo motor works of Portello, on a property of about 230000 m^2. The scheme consisted of two new exhibition pavilions, forming a natural northwesterly extension of the existing facilities of the Milan Trade Fair. In addition, on the side of the property towards the ringroad in Milan, a multifunctional complex was planned, comprising a centre for business communications, office buildings, hotels, service flats and retail premises. A new congress centre was designed to be the pivot between the existing Fair and the business centre. This was to be built by the developer for the Commune of Milan in return for the municipality-owned land and its associated building rights. In fact, the commune, together with *Ente Fiera di Milano* and *Sistemi Urbani*, were the joint owners of the project site. *Ente Fiera di Milano* is the company that owns and runs the Milan Trade Fair, whereas *Sistemi Urbani* forms part of the public holding *IRI–ITALSTAT*, which for a few years starting at the end of the 1980s grouped all the companies in the real-estate sector in the subholding *IRITECNA* (now liquidated).

The project for the new Portello involved the redesign of the road plan and the creation of a new underpass, to rationalize both private and heavy goods traffic to the Fair. The ultimate aim was to reduce the high levels of traffic within the city as a whole. The project called for the creation of private and public parking facilities for over 9000 cars. Green spaces and pedestrian areas were carefully positioned, according to the developer, to integrate the new development into the existing urban context.

The area affected by the detailed plan lies in the northwest of the city. It is bounded on its inner and outer limits by the middle and outer ringroads and

Table 6.2 Detailed project data of the new Portello.

Site area	230000 m^2
Site owners:	
Commune of Milan	32%
Ente Fiera di Milano	22%
Sistemi Urbani SpA	46%
Gross floorspace	268000 m^2
Total building volume	1300000 m^2
Public green spaces	61000 m^2
Public parking	125000 m^2
Private parking	80000 m^2
Floor area classified by function:	
Congress centre	22000 m^2
RAI-TV centre	26000 m^2
Exhibition facilities	85000 m^2
Offices	93000 m^2
Retail premises	10000 m^2
Hotels/service flats	32000 m^2

Source: Sistemi Urbani.

radially by the Corso Sempione–Via Certosa and Corso Vercelli–Viale Scarampo axes. The area is marked by the contrast between the premises of the existing Fair, which are laid out on a north–south plan, and the surrounding residential area that has a southeast–northwest orientation. It is along this axis that the huge former Alfa Romeo works lie, on an extended rectangular site interrupted only by the outer ringroad.

This area and the adjacent Fair constitute an obstacle, resulting in a serious lack of continuity between the central part of the city and the outskirts. The site lies across an area in which the densely built blocks of the centre give way to the more dispersed structures of the suburbs, punctuated by large-scale sports facilities (such as the San Siro football stadium), green recreational areas (Monte Stella) and large open spaces.

The project site was divided into three principal areas, which are briefly described here. The first area comprises the extension to the Fair facilities, which involves the building of pavilions along the entire length of Via Scarampo, to link the existing Fair at one end and the congress centre at the other. This building was planned on the corner of Viale Scarampo and Viale Serra, to form the pivot of the whole project, surrounded by 28000m^2 of public car parking. The exhibition pavilions had two levels with interior galleries, making up a planned total of 85000m^2. It was planned to provide a further 48000m^2 of public parking and 17000m^2 of private parking space. Along Viale Scarampo it was envisaged that an additional 37000m^2 of parking space would be distributed among an extensive landscaped green areas (Fig. 6.6).

The second project area was assigned to radio and television facilities and associated private parking space. This building was to be located at the extreme edge of the site alongside Via Traiano, where it would have balanced the mass of the congress centre on the other side of the site next to the ringroad. The third area, entirely under the responsibility of *Sistemi Urbani*, would contain the office, hotel and shop facilities, divided into two separate clusters. The first was located between Via Traiano and Via Gattamelata, and composed 50000m^2 of office space as well private parking facilities. The second, situated at the centre of the area overlooking Viale Serra, was distributed around the large square in the centre of the complex. On the roadside, the office twin towers, the project's most prominent buildings, were planned (20 storeys high with 42000m^2 of floorspace). On the other side of the square, the project located the hotels and service flats as well as the large shopping centre, plus parking spaces of about 13000m^2.

The Portello project grouped together in one area the typical range of building types normally found scattered in the whole property market. In order to harmonize with its surroundings, the layout prescribed a system of open spaces, building blocks and pathways, in various forms. Hence, the developer was rather confident that the project would have been a success. It must be remembered that, at the end of the 1980s, the real estate market of Milan was in a real boom period. For example, between the spring of 1989 and the autumn of 1990, prices for new property in all categories rose between 20% and 30%, whereas those for renovated property showed rises of between 15% and 20% from an already high base level.

Plan implementation

The Portello project was a complex venture, not just technically or financially, but also in terms of site ownership. Before being bought by *FIAT*, Alfa Romeo passed the Portello property to *Finmeccanica*, one head company of the public holding *IRI*. *Finmeccanica* transferred the property to *Sistemi Urbani*, a company controlled by *IRI* through *IRITECNA*, the same subholding that conceived, through *MEDEDIL*, the Naples business district project. Thus, when the project was conceived, the area was owned by three major actors: *Sistemi Urbani* (46% of the site), Commune of Milan (32%) and *Ente Fiera di Milano* (22%). At the end of the development, the expected ownership was planned as follows: Commune of Milan (congress centre), *Ente Fiera di Milano* (exhibition pavilions), *RAI* (television production centre), *Sistemi Urbani* (offices, hotels, service flats, shopping centre).

According to the promoter, the Portello redevelopment plan (drawn up by *Sistemi Urbani* and *Ente Fiera* for their respective parts of the scheme) was in perfect concordance with the redevelopment objectives expressed by the Commune of Milan in the city master plan and its updates. In July 1989 the city council approved the detailed plan for the Portello project, whose proposals were stated as follows:

- to emphasize the importance of the Milan Trade Fair in the national economy and to reflect the increasing demand for permanent facilities for business communications and exchange
- given the decision to maintain the Fair premises in their central location, to create a framework within which the facilities of the Fair could be extended and new facilities for business communications created
- to enable the city to evolve a more complex structure suited to its role as the main Italian centre for business communications and exchanges, a function supported by the proposed congress centre
- to provide hotel, office and shopping space to service the exhibition and meeting spaces
- to create parking space for the users of the above facilities, and for the heavy goods traffic associated with the exhibition facilities
- to complete the road network, extending Via Gattamelata across the whole of the site in order to separate user vehicles from general traffic, and to improve circulation in the whole area.

The project was expected to enter the construction phase in 1991 and to be completed by the end of 1995. As of 1996 the project is far from being completed as initially planned and the chances are dim that it will ever be. Basically, just the new pavilions of the Fair have been built, whereas the office towers had to be dropped, because the Commune of Milan refused to authorize them. This represents a major change compared with the optimistic view of the developer up to 1990, when the sale of office space, retail and accommodation facilities was expected by *Sistemi Urbani* to be completed before the end of 1991.

Indeed, as of 1991, the placement plan for the Portello complex envisaged two alternative modes of sale: the sale of serviced lots or the sale of finished buildings. By the first method, which was considered more desirable and believed to be used in the majority of cases, the purchaser was responsible for the construction of the building while the area services (the road system, utilities, parking and landscaping) were to be provided by *Sistemi Urbani*, the project promoter. The cost of the serviced lots was set as follows (prices at 1990 levels): 2 million lire per m^2 of floorspace, equivalent to 0.6 million lire per m^3 of building volume. The exhibition and television production facilities were not meant for sale. The project was to be almost entirely financed by internal funds and proceeds from advance sales.

In March 1990 an agreement was signed between the Commune of Milan and *Sistemi Urbani*, to regulate the exchange of the land owned by the city and its associated building rights, as well as the overall planning permit procedures, including the anticipated construction of buildings for the Milan Trade Fair. It must be noted that between May and October 1989 (that is, before the mentioned agreement with the municipality), the promoter had signed pre-sale contracts with prominent clients (*Bayer SpA*; *Leblon SrL*; *Milano Sviluppo Portello SrL*) for the bulk of the office space, including the two office towers that would have become the emblem of the project.

But in July 1992 a serious conflict between *Sistemi Urbani* and the Commune of Milan was made public. In fact, the promoter recurred to the law courts, accusing the municipality of delay in the granting of planning permits without reasons; according to the promoter they were to be issued with the enacting of the agreement made in March 1990 and ratified by the mayor in March 1992. Over time the legal controversy became more and more intricate, with innumerable sentences at all levels of the Italian judiciary system (first level court, *TAR*, the regional administrative court, *CORECO*, the regional controlling institution, *Consiglio di Stato*, the highest consultancy body for administrative and legal procedures of public administration).

As the legal dispute went on, the promoter could not abide by the conditions of the pre-sale contracts, and the advance payments of potential clients of the Portello project had to be reimbursed plus interest. In January 1995 the *TAR* of Lombardy accepted the arguments of the promoter and ordered the Commune of Milan to examine within 60 days the applications for planning permission from *Sistemi Urbani* concerning the twin towers and other office and commercial spaces. On the one hand, the commune opposed the decision of *TAR* in front of *Consiglio di Stato*, and, on the other hand, did examine the applications in question. The outcome was a rejection of all the applications, on the basis of the following arguments: the requests were found incomplete on administrative grounds; there was a widespread opposition to the project in the city; and there had been a negative vote of the elected neighbourhood council (*Consiglio di Zona*), whose opinion is not binding for the city council but has a certain weight, especially in controversial cases; and the applications of the developer were deemed not in line with the city planning standards applicable for the Portello area according to the city master plan.

In the meantime, the Milan Trade Fair went ahead to utilize the land and the building rights it had received from *Sistemi Urbani* as part of the initial agreement, in spite of the legal opposition of the promoter. The Fair development plans were supported by the Commune of Milan. A specific *Accordo di Programma*, an agreement regulated by the law of 1990, was signed in May 1994. It defined the policies of the key local institutions concerning the long-debated issue of the development of the Milan Trade Fair. It was decided not to transfer the Fair outside the city, as some had proposed, but to support its expansion in the same area (also on the land owned by the municipality) and in the framework of the global reorganization of the Portello zone. These policy objectives were translated in

July 1994 into a new variant (Z18) of the city master plan for the Portello area. The new variant meant the official end of the previous detailed plan and hence of the development projects based on that plan. In addition, in October 1994, the Fair reversed to *Sistemi Urbani* all responsibility for the soil pollution found by the local public health institution (*USSL* 75) in the property previously utilized by the Alfa Romeo motor works and then transferred to the Fair by the promoter.

In June 1995 the Commune of Milan defined a programme of upgrading for several large urban districts, in line with the agreement of May 1994 and with a ministerial decree of December 1994, which had required certain large cities to initiate redevelopment actions. The new programme included the Portello area, as planned in the new variant Z18 of the city master plan, plus the zone north of Portello and some other areas owned by *FIAT*. In October 1995 two new promoters, *Auredia* and *PALEO* (the latter of the *FIAT* group), applied for planning permission to develop offices and a mixed-used complex (for about 92000 m^2 of gross floorspace) on the land owned by the municipality and by the Fair in the Portello area. In exchange, the new promoters would build free of charge for the municipality (to pay for the *oneri di urbanizzazione* and for the building rights of the municipality) a public park of 10 ha, carparking spaces for the Fair, and the tunnel to put underground via Gattamelata, which is the real bottleneck in the traffic flows in the area.

As of the beginning of 1996, the initial promoter, *Sistemi Urbani*, is still pursuing its legal battle against the municipality to obtain a financial reimbursement for the delays in the municipal decision-making process, and for the final refusal of planning permissions according to the original Portello project.

Conclusions

This is a typical case of lose–lose negotiation. In fact, it is not possible to identify a "winner" in this case, but it is certainly possible to list the losers: basically all the actors directly involved, and the city at large. Starting from 1992 the development became more and more an intricate legal affair with the only benefit accruing to the lawyers involved, and at a cost to the Commune of Milan, *Sistemi Urbani*, the Milan Trade Fair, and with damage to the local community and to the city as a whole.

From 1992 it was no longer a planning case. It became a case of planning law, good only for a reference manual of planning practices to be avoided. It appears that, from the viewpoint of constructive urban planning, a major mistake was made by the promoter, who thought that a solution to the apparent stalemate could be reached in the law courts. First, in the right circumstances, the Italian judiciary system can oblige a commune to examine (or re-examine) an application for a planning permission, as *Sistemi Urbani* obtained from *TAR*. But in Italy no-one can impose on a commune, whether as large as Milan or as small as Ferrara, the obligation to examine and to accept any application if it conflicts with the existing city master plan. Secondly, the communes have the enormous power to approve the city master plan and change it at will, according to the perceived general interest.

In the final analysis, it is the master plan that orchestrates the development of the public infrastructures that affect accessibility to the urban territory, and it is again the city master plan that determines the different uses of each portion of urban land and also the volumes of construction allowed on each portion. In sum, it is the city master plan that determines the urban land rent values and the consequent windfalls. As the plan is decided by the city administration, that administration ultimately has the power to decide on the levels of urban land rent, which are the cornerstone of the fortunes of all real estate promoters. In conclusion, through the master plan the city administration can decide the fate of the property promoters. Against such a tremendous power, any major legal battle leaves only losers behind and should be avoided.

A mistake was also made by the city administration, for it acted until mid-1994 with no real policy on the underlying key question, that is whether or not to relocate the Milan Trade Fair and, consequently, what to do with the Portello area. It may seem a paradox, but the city administration was forced to elaborate an urban policy, and consequently a new variant to the city master plan, by the legal action of the promoter. The commune played a rather passive role throughout the case, perhaps through fear of making some mistakes. This can be explained by the climate of those years, when in the whole country local public administrators (especially those who had belonged to one of the corrupt government parties) were everyday the target of sharp criticism by the media. The non-decision was a form of defence, albeit not consistent with any policy of urban renovation and social growth.

In conclusion, the Portello project, as a whole, has not been a success, because it has failed to develop the old industrial area into a mixed-use complex as initially planned. Given the presence of just three major actors (Commune of Milan, *Ente Fiera di Milano* and *Sistemi Urbani*), the implementation process could have been conducted more smoothly. Looking back, it seems that no one really gained from the outcome of the project. Even the Milan Trade Fair, which managed to build part of its facilities, has been hurt, because the accessibility to the Fair has not been sufficiently improved.

From the beginning, it was observed that the Commune of Milan had high stakes in that venture, since it owned directly one third (32%) of the land, and could strongly influence the decisions of the "autonomous institution", *Ente Fiera di Milano*, which owned 22%, making a total of 54%. In addition, through the city master plan the commune could determine permitted development on that site (as on any other site within the city boundaries). In brief, the fortunes of the project were determined by one of the actors who was directly involved in the recycling of the site. Nevertheless, it seemed from a financial viewpoint that the initially planned outcome of the development would benefit the Commune of Milan only marginally, whereas the bulk of the profit-making project (offices, hotels and commercial space) would have been taken by *Sistemi Urbani*. By contrast, in the case of the Naples business district the key public functions (city headquarters of the Italian Mail Company and the Law Courts) have been located on the half of the project land owned by *MEDEDIL* (controlled, like *Sistemi Urbani*, by *IRI–ITALSTAT*), and this became the reason for success of the project, whereas on the other half of the land, owned by the local authority, nothing happened.

It is plausible that the changing social climate of the early 1990s in Italy had a crucial influence on the difficult relations between the actors involved. The Portello project spans a period of time through which the economic and social conditions of Italy changed significantly. Starting in 1992 the Milan real estate market was affected by two major factors: the worsening of the property slump and, notably, the spreading of public inquires into the national business world (known as *Mani Pulite*). The latter factor led the local institutions to stiffen their normal procedures for granting planning permissions. But it also marked the complete change in the role of the state-controlled companies. In fact the *partecipazioni statali* system meant the direct control of the ruling political parties (first of all, the Christian Democrats and the Socialists) on the decision-making of the companies owned and financed by the state. This condition was at the base of the corrupt system uncovered and dismantled by the judiciary, starting in 1992. The phenomenon was a true revolution in Italian civil society against the cancer of the corrupted parties, institutions, and men.

6.5 Genoa: the Baluardo project in the framework of the revitalization of the medieval port

Context of the development

Genoa is the most important Italian port. The city was one part of the so-called "industrial triangle", together with Turin and Milan, during the Italian economic boom of the 1950s and 1960s. The city suffered from the competition of northern European ports in the 1970s and part of the 1980s, for several reasons, including: the introduction of the worldwide container freight system, which would have required investments and new technologies that lagged behind in the city in that period; the poor connections of Genoa with the rest of Europe; and the lack of response from public and private actors to the changing economic needs of the international business community.

In brief, the city suffered for the same factors that affected other traditional industrial cities, with the added element that Genoa had a severe shortage of land for its development. Hence, any new industrial site that was made available for re-use was seen as a true opportunity for the whole city. It was in this spirit that the city administrations of the 1980s prepared the programme for 1992, 500th anniversary of the discovery of America made by Christopher Columbus (*Cristoforo Colombo*), the most celebrated citizen of Genoa. The celebration programme took the name of *Colombiadi* (the Columbus Fair) and was intended as a long-awaited event to re-launch the image of the city to the world.

The most visible part of the *Colombiadi* programme was the restructuring of the medieval port of the city, facing the historical centre and close to the central square of the city (Piazza de Ferrari), plus a series of public works to ameliorate urban traffic conditions. Freight ships had long before the 1980s ceased to use the old port, as new and much larger infrastructures had been built over the years elsewhere in Genoa and also in other communes along the coast. But the old port had continued to be a separate body from the city, being cut off by a high iron fence (the internal area was restricted to "custom services") and guarded gates. The project aimed to open up that area to new uses (aquarium, exhibition centre, congress centre, retail and office activities) and to give way to the revitalization of the waterfront. It was perceived that the city should try to diversify its economic base and no longer be solely dependent on a traditional port economy. That was the plan or, rather, the initial hope. The practice proved to be different.

Recently released data show that a total of 530 billion lire of public money was spent on the revitalization of the port, out of a total available financing from state funds of 585 billion lire. The restored buildings have a gross floorspace of 70000 m^2, for an estimated value of 280 billion lire. These buildings, previously owned by the state, have passed into the ownership of the Commune of Genoa for just 400 million lire, thanks to a law passed in 1993. A bargain for the city? Not quite. The commune found overwhelming difficulty in identifying suitable activities for the properties (which, obviously, could not be resold), whereas the maintenance costs are anything but negligible. According to data released in December 1995, in the period 1992–5 the vacancy of the restored buildings resulted in 90 billion lire of lost income, plus an additional 15 billion lire of maintenance costs.

A recent, albeit late, move in the right direction are the plans of *Porto Antico*, a joint venture between the Commune of Genoa (80% of the operating capital) and the Chamber of Commerce (20%), announced in December 1995. The company has taken over the role of *Ente Colombo*, the special institution set up to implement the Colombus Fair, enlarging its mission to coordinate the revitalization of the area and induce inward investments. According to *Porto Antico*, the port area is likely to attract 2.5 million visitors in 1996, thanks to the planned initiatives of the aquarium, the leisure port (250 boats), and a new series of conferences and entertainment activities. Overall the port revitalization programme would generate 400 new jobs per year. Again, the intentions may be the right ones, but there is little room for new mistakes given that the city lost prominent potential investors because of a lack of coordination between the public authorities in charge, for one reason or another, of the port.

For example, the *Magazzino dei cotoni* (cotton warehouse), the largest building in the port (three storeys, one single block, 0.5km long) is still vacant, although the structure, which was completely rebuilt by the renowned architect Renzo Piano, seems suitable for a variety of profit-making uses. The building was renovated to be used as a trade fair centre, but since the inauguration period of the *Colombiadi*, it has been completely vacant. The main reason why this symbolic building has not been utilized for fairs or exhibitions is not the weakness of the city in organizing public meetings and commercial events, but rather the lack of coordination between all the public bodies involved in the decision-making process.

In December 1994 the British company UCI (Paramount and Universal Picture) proposed to utilize 15000 m^2 of the *Magazzino dei cotoni* as a multi-hall cinema

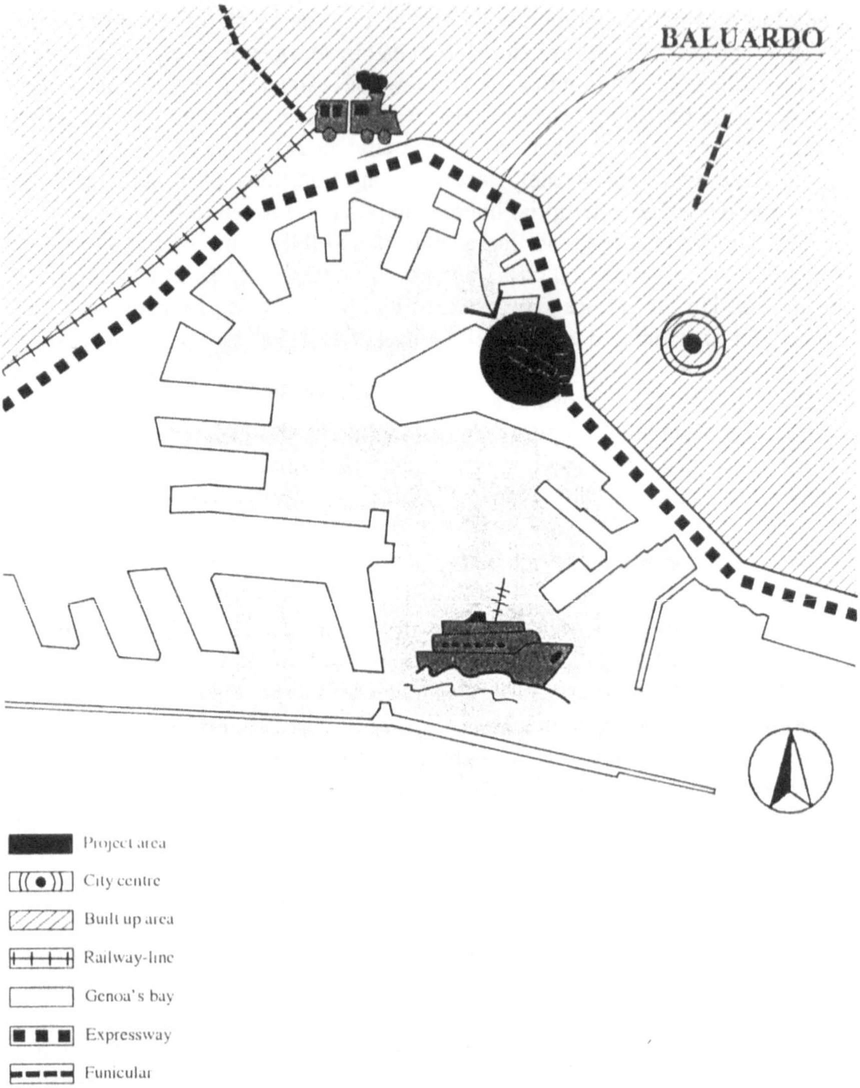

Figure 6.7 Location of the Baluardo project in the medieval port of Genoa.

plus other entertainment activities. The potential investment amounted to 10 billion lire, plus an annual rent of 250000 lire per m^2 and a further 100000 lire per m^2 per year as contribution for the maintenance costs of the premises. It was estimated that the multi-theatre would have acted as a magnet for about a million visitors per year, with obvious spin-offs for the whole area. That proposal was vetoed by the *Sovrintendenza*, the state institution in charge of preserving the listed monuments, including the examples of industrial archaeology such as the *Magazzino dei cotoni*. It seems that the placement of the multi-hall cinema inside the building would have required some changes to the internal distribution of space. The problem is that almost any activity would require some sort of internal changes in the building.Given the briefly outlined framework, it comes as no surprise that the Baluardo project had a hard life.

The content of the plan

Baluardo is the name of the seventeenth-century fortress beautifully located just outside the area renovated for the Colombus Fair, but inside the medieval port. The building had been used for military and police purposes. In the 1950s the property was occupied by artisans and retail business, then it was left vacant. From the top floor the view of the port and of old Genoa is unforgettable. The building has five storeys and a gross floor area of about 6000 m^2.

As the fortress is a listed monument (according to the law of 1939), the restoration project had to abide by the strict rules and specifications prescribed by the *Sovrintendenza*, including the use of approved construction materials (e.g. natural stone cut by hand and forged iron) and frequent inspections on the construction site by technicians of that controlling institution. The project aimed to free the building from some spurious volumes added over the years. Although housing demand was strong in Genoa at the time, the property could not be converted into flats, as this would have required demolition works that would not have been authorized. The final utilization was then office use, with the exception of the ground floor, which was found more suitable for retail activities. At the request of the *Sovrintendenza*, the promoter had to restore the antique guardian walkway on the top of the fortress, and open it free of charge to the general public according to an agreed timetable.

The location of the project is good but accessibility is poor, because the property has been cut off from direct access to the new and adjacent public parking zone built for the Colombus Fair in the old port (see Figure 6.7). The promoter asked the *Ente Colombo* to build a passageway for pedestrians between the fortress and the new parking zone, on the premise that, according to the promoter, the fortress had always had a direct passage to that section of the port. Obviously, the market value of the property would have been enhanced by any measure that would have improved its accessibility to a modern and guarded parking area. In fact, in the whole of central Genoa, traffic is congested at all hours, and the availability of parking space is a rare plus in the real estate market. Apparently, the promoter expected the public authority to bear all the costs of that infrastructure, and this caused a stalemate. According to a different source, the public authority did not consider the construction of the pedestrian passage an infrastructure having a general public utility, regardless of cost sharing. As of mid-1996, the passage had not been built.

Plan implementation

Initially, the project was a joint venture between *Fortune SpA*, a real estate promoter based in Genoa, and *Gabetti Holding*, the sole real estate trade group with a national network in Italy. In 1989 Gabetti took over the operation completely, acquiring full control of *Baluardo Srl*, the project operating company named after the fortress.

The promoter (*Baluardo Srl*) applied for planning permission in 1985, well ahead of the Colombus Fair planned for 1992. Having obtained the necessary green light from the *Sovrintendenza*, the Commune of Genoa granted planning permission in March 1987. Construction works began in February 1988. In March of 1990 the promoter requested permission of *variante in corso d'opera*, a change of the initially approved project. This application was accepted by the *Sovrintendenza* in March 1990, and the commune granted the permission in May 1991. The certificate of completion of the restoration works was issued in June 1993.

The implementation period spanned eight years for a project rather small in size, but requiring a double turn of public authorizations (the Commune of Genoa and the *Sovrintendenza*), which are usually granted in sequence and not in parallel. In fact, it is a standard procedure for a commune to await first the response from the *Sovrintendenza*, and only after its favourable response to examine the application for a planning permission (or a change to an already granted permission). Hence, the approval process takes about twice the normal time, when a building is a listed monument. When the two institutions (commune and *Sovrintendenza*) have different views, as in the case at hand, the requested time can be even longer.

The project was financed by the developer at a cost of 14 billion lire. The expected gross profit of the operation was initially estimated at 4 billion lire. However, the marginal rate of return of the intervention is being gradually lowered by growing financial costs. According to the promoter, the placement of the property in the real estate market is lagging behind schedule, because the underused facilities of the renovated antique port have failed to create a favourable economic climate for the area.

Conclusions

The project was a success from the viewpoint of the technical renovation of an antique fortress, a listed monument requiring special care. The construction phase was completed on time and within the allocated budget. The restoration was conducted with the agreement of the *Sovrintendenza*, and the construction also gave the opportunity to free the public space surrounding the building of some obsolete constructions that were a visual blight for the port visitors. However, the Baluardo project did not generate the expected profit, because three years after its completion date a good part of it is still unsold and unlet. According to the initial expectations, the project may be regarded as a failure.

This is also true if we refer to the whole story of the Colombus Fair. Here the lesson is remarkable. First, it should be clear that the revitalization of an inner area cannot be tackled solely through an architectural project. What is needed is a global vision of the desirable direction of growth (not necessarily physical growth) of a city. Further, it is also necessary that public and private actors cooperate and actively work towards a common goal: to stop the downturn of the city. Secondly, from a planning viewpoint, it should be crystal clear that the action of urban renovation calls, as a prerequisite, for the establishment of a kind of "economic development agency" with the participation of all the major local institutions (commune, province, region, chamber of commerce), plus the key private enterprises and economic associations. More and more there needs to be true cooperation between all local and state institutions that have localized jurisdictions.

It seems that the promoter is completely right in pointing out the inefficient handling of the Colombus Fair. It is also true that the failure of the Fair caused economic harm to the Baluardo project, as it increased the time required for the letting and sale of the project, thus increasing the debt servicing costs, and finally reducing the profits expected at the end of the operation. Nevertheless, one should also point out that, in this case, the promoter took the risk of locking the fortunes of the project with the outcome of a more global project whose decision process was obviously out of its control. It was a real estate venture with a component of gambling, rather than pure economic planning.

If the Colombus Fair, which was financed entirely through taxpayers' money, had been successful then it would have brought to the project windfall earnings significantly higher (for comparable investments) than in other parts of Genoa. One may wonder whether all the private promoters who would have made high profits as a consequence of that success would have made public their praises for the activities of the local institutions, or would have remained silent. In conclusion, it appears that the dialogue between public and private actors is something that is worth pursuing in advance of any major investments. The absence of cooperation can lead to failure of any urban revitalization programme, and can have serious financial implications for the private developer.

6.5 Out-of-town development: the case of the shopping centre "*Le Fornaci*" in Turin

Context of the development

This is the story of the construction of a shopping centre in the outskirts of Turin. The description will highlight the basis of the discourse underlying the retail sector in Italy, which since the mid-1980s has been undergoing structural change. Whereas in Great Britain and France the market is mature (and there are frequent signs of saturation), in Italy the fastest rising property market sector is the large-scale shopping centres, which are mainly built at out-of-town locations or in vacant industrial areas. Another distinctive character is the fact that in Italy for each shopping centre there are two authorization processes, one for granting the planning permits for the physical construction of the centre, and the second for the permits to open and operate the shopping centre as a retail business activity of various merchandise categories. All the above should make clear, I think, why it is necessary to analyze in detail an ordinary case of a shopping centre, in order to explain the distinctive characteristics of the Italian retail property market.

Turin, the third largest city of Italy after Rome and Milan, is renowned as the industrial capital of the country, and as a leading world centre of businesses related to car design and car manufacturing. Names such as FIAT, Lancia, Pininfarina and Giugiaro, to mention just a few, are associated by the general public with the name of Turin. Nevertheless, the city is also known for its unique natural setting, on the edge of Padana Valley with views of the Western Alps, and for its urban centre of well preserved Baroque-style architecture. Turin is the most important centre for the telecommunications industry: both RAI, the Italian state television, and *Telecom Italia* were founded there, together with the Italian movie industry. The Olivetti company, located in the area, has recently entered the mobile phones business. In the city one can also find the leading national firms of the banking system (Bank San Paolo) and of the textile industry (GFT Group).

Recently the city has increased its international standing. For example, in 1993 the European Union decided to locate in Turin the Agency for Training East European Managers. The city has housed since the 1960s the training campus of the International Labour Office and, according to Boutros Ghali (General Secretary of the United Nations), Turin has a good chance of being selected as the seat of the training college of the military forces of the United Nations. In 1996 the city was chosen by the European Union as the seat of the Intergovernment Conference for the revision of the Maastricht Treaty.

The facts briefly outlined above should testify that the image and the economic base of the city are shifting gradually from a pure industrial city to a metropolis with a diversified economy. However, the city retained a rather traditional retail sector, with thousands of small stores. In fact, the single economic branch where Turin was known for it being less modernized than other major Italian cities is the retail sector. Since the mid-1980s, the area has been the target of both national and international groups, including *Euromercato*, *Città Mercato*, PAM, *Auchan* and *Continente*.

The prevailing form of retail distribution in Italy is the traditional store. This is highly specialized, because the commercial licences are granted by the communes according to a great number of *Tabelle*, that is of genre of merchandise for retail activity. For example, a bakery cannot sell fresh milk products, whereas the general foodstore can sell frozen bread but not fresh bread, newspapers can be sold only by news-stands, and so on. The traditional stores are usually very small (30–40 m^2) and are owned and run on a family basis. Opening hours are strictly regulated by the commerce department of the commune, and they abide to strict rules deep-rooted in history. For example, in many cities, barber shops are closed on Mondays, food stores on Wednesday afternoons, and butchers close on Thursday afternoons, and so on. There is usually a long break between 12.30 pm and 15.30 pm, or even later in most cities. Exceptions to these general rules are granted, on a case by case basis, for holiday areas (but only during the holiday season) and in the period of 2–3 weeks before Christmas.

The open-air markets are still found in most cities and have been part of the urban scene for centuries. For example, in most cities one can always find a *Piazza del Mercato* or *Piazza delle Erbe* (market square) close by the local city hall and the main church. Indeed, the *piazza* has been the distinctive character of the Italian architecture since the Middle Ages, throughout the Renaissance. The *piazza* has always been the meeting place of any Italian city and, at the same time, the place to conduct business negotiations, or to buy agricultural produce and miscellaneous goods, or to hold public meetings and concerts. Given this background, one can understand why the large-scale retail centres are seen as alien elements, forerunners of a model of the city (and of life) typical of North America, but with no background in the urban and architectural traditions of Italy.

Between 1972 and 1992 the region of Piedmont (an area of about 5 million people) granted 13 permits concerning full-scale shopping centres, 119 permits for supermarkets (food-only large retail centres), and 61 shopping centres without an internal food market. These data refer to new retail centres (50%) and modifications to existing businesses (50%). Since the early 1990s the fastest rising category of retail centres, in Piedmont as well as in the whole North of Italy, are the "mini-markets", stores with less than 200 m^2 of floorspace affiliated in franchising to a discount chain. This kind of store does not need approval at regional level, but only a permit from the commune concerned. In addition, there is not the need for any planning permit for construction, as the mini-markets usually take the place of two or three adjacent traditional stores pushed out of business by the large retail centres. It seems that this could be a form of retailing that blends the tradition of the small Italian stores located inside the city with windows facing the streets, with the quality and low prices that are typical of national chains.

The content of the plan

The project is located at the southwestern boundaries of Turin in the small Commune of Beinasco. The location is extremely good because it has a catchment area of several hundred thousand people, encompassing small cities such as Orbassano, Rivoli and Nichelino. The centre also attracts clients from other fringe areas of southern Turin, because it is close to one exit of the ring motorway around the city.

Prior to this project, the land was mainly classified for "industrial use" in the master plan of the Commune of Beinasco. In July 1982 *Coop Piemonte*, although not yet owner of the land, advanced to the Commune of Beinasco a request to open a shopping centre of 6000 m^2 of floorspace in the area known as *Le Fornaci*. The commission of retail commerce of the commune examined the request and approved it in April 1993, subject to the signature of a *convenzione*, an agreement, between the promoter and the commune itself. In April 1983 the land was still classified as "industrial", so three months later, in order to make the project consistent with the city master plan, the Commune of Beinasco approved the *variante 3* (change 3) to the said plan. This act was justified on the basis that the requested project, already approved by the retail commission of the commune, was deemed consistent (by the commune itself) with the planning objectives of the superior local government bodies (the *Comprensorio* of Turin and the region of Piedmont) in the domain of location of large shopping centres. However, as will be explained in detail below, these institutions had not yet expressed any official evaluation on the project.

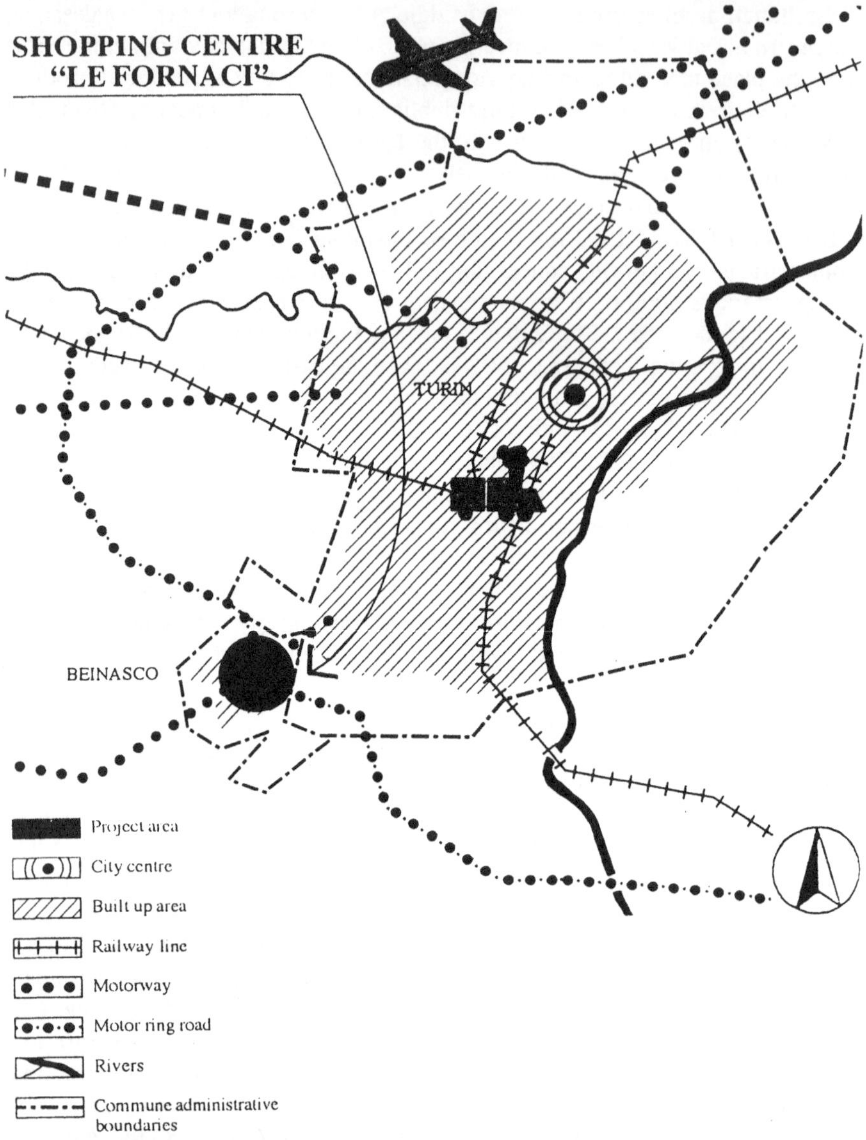

Figure 6.8 Location of the shopping centre "*Le Fornaci*" in Turin.

The project land was codified in three different classes: Ia2.1, Ib2.1 and a third area on which it the *Drosso* exit of the ring motorway was planned. Class Ia2.1 was still industrial use, and the class Ib2.1, covering the best part of the property, prescribed retail use and related activities. In this area, the issue of the planning permit from the Commune of Beinasco was subject to the approval of a *Piano particolareggiato* (*PP*), a detailed plan consistent with the city master plan, and to the issue by the region of Piedmont of the *nullaosta*, the authorization to open a shopping centre with a maximum floorspace of 20000 m^2.

The land of 6.6 ha was acquired by *CIMIN–SRL* from *Edilopera SAS* for 2.8 billion lire (or 42000 lire per m^2) on 31 July 1985. About 2.5 ha of land was passed to the Commune of Beinasco for the construction of public services (including the motorway exit), according to a previous agreement between the commune and the vendor. The price paid for the project land may seem low, but at the time it was higher than the price of comparable sites that were zoned for industrial use or agricultural use. As prescribed by the city master plan, the issue of the planning permit was made conditional upon the signature of an agreement (*convenzione*) between the promoter and the Commune of Beinasco. The document (dated 23 January 1987) required the promoter to build on behalf of the Commune of Beinasco public works for the amount of 2.9 billion lire, plus a cash payment of 625 million lire to equal the required amount of 3.5 billion lire as the sum due for the *oneri di urbanizzazione* (the tax for the planning permit), calculated according to the type and size of the authorized construction. In fact, three days later than the above-mentioned signature, planning permit 56/86 was granted by the Commune of Beinasco. Later the promoter requested and obtained from the commune two additional planning permits (the first one in February 1988, and the second one in June 1988, the same month as the completion of construction works) to change the internal distribution of floorspace, within the already approved total floorspace.

In parallel with the authorization process for the physical construction, the promoter sought the permits for the beginning of the commercial activities of the centre. In March 1984 the department of commerce of the Piedmont region authorized the opening of a shopping centre of 5950 m^2 of floorspace, limited to *Tabella VIII*, that is all food items. The permit for a further 2050 m^2 was granted in February 1987, for the sale of all other types of merchandise (*Tabelle* from IX to XIV). As a result of the step-by-step authorization process, briefly described above, the final result was a fully fledged shopping centre, and not a shopping facility at neighbourhood level as envisaged by the *Comprensorio* of Turin in 1983, the public body that in the meantime had been abolished. So, the shopping centre in Beinasco reached 8000 m^2, as did the one in the adjacent city of Rivalta, which was supposed to be the only one in line with the planning objectives of the *Comprensorio* of Turin, as it is clarified below. The threshold of 8000 m^2 is the size of floorspace, which is usually referred to as "a shopping centre", and which requires authorization at regional level. To this area, the promoter of the centre of Beinasco, as it is usual, added other satellite retail businesses, whose floorspace was not subject to the permission of the region of Piedmont.

Overall, the project gross land surface is 40918 m^2, of which 14176 m^2 are occupied by the buildings. The gross floorspace is 19523 m^2. The main building has three floors above ground. The ground floor is the bulk of the project with 14176 m^2 of floorspace. At this level there is the shopping centre, which comprises a large open surface (8000 m^2 as described above) for non-specialized shopping, a public mall with showrooms and artisans, and a covered street with specialized small stores. The first floor (3707 m^2) and the second floor (1642 m^2) are basically used as office space. The layout includes, outside the main building, a parking area of $19932\,m^2$, a *piazza* of $1599\,m^2$, pedestrian areas and green spaces to a total of $3053\,m^2$.

Plan implementation

When the idea of a shopping centre in Beinasco was conceived, there was an intermediate local government body (named *Comprensorio*) between the Commune of Turin and the Province of Turin. In fact, in Italy in the 1970s and early 1980s, most planners thought that urban and regional planning was not efficient, because of a strict subdivision of powers between the traditional local government bodies (region, province, commune) and because the boundaries of the lower government tiers (commune and province) no longer matched what was thought to be the actual distribution of homogeneous economic zones in the region. Much time and taxpayers' money were wasted in debates on the criteria to be used for the identification of the correct boundaries. At last, a new government body was set up. In most cases it covered a geographical area larger than the communes head of province and smaller than the related province. Obviously, the new local body had a short life, but at the time of the project at hand it was in full swing, and its planning power included the large-scale retail sector. In this domain, the new body could issue recommendations that were not binding for the communes but had, at the time, a certain political weight and could determine the final decision, which was the domain of the region of Piedmont. The plan of the *Comprensorio* of Turin envisaged only one centre of $8000\,m^2$ of gross floor area in the southwestern part of the metropolitan area of Turin.

The proposed project in Beinasco ($6000\,m^2$ of gross floorspace for the actual shopping area), was examined by the *Comprensorio* in parallel with a similar request ($8000\,m^2$ of gross floorspace for actual shopping area) proposed by the group *Euromercato* for a centre in Rivalta, an adjacent small city. In the recommendations issued on 14 December 1983, the *Comprensorio* maintained that the centre *Euromercato* was acceptable, because it could have equilibrated the presence of a large shopping centre already built in the northeast of Turin. On the contrary, the Beinasco project was not consistent with the objectives of the Plan of *Comprensorio*. In addition, the centre of Beinasco was deemed located in a zone that would not have generated polarization effects, and would have had a mainly local attraction. As a consequence, both cases were approved: the centre in Rivalta as a full shopping centre of metropolitan level, whereas the Beinasco centre was recommended only as a shopping centre at neighbourhood level. In practice, however, the green light was given to both. Once built, the centre of Beinasco was expanded gradually to the same level of that in Rivalta, as a consequence of several additional planning permits.

Paradoxically, the Commune of Beinasco granted the initial planning permit also as a form of protection of its urban environment. In fact, in case of rejection, the proposed shopping centre planned in the adjacent city of Rivalta would have attained, if approved and built (as it happened in the late 1980s), a greater role in the retail market of southern Turin, thus attracting an increased flow of private traffic from Turin, which would have crossed Beinasco anyhow. So, in the end, the Commune of Beinasco would have only borne the environmental costs and received none of the benefits of the change of the retail market. From this viewpoint, this case shows that the urban policy in the field of retail and large shopping centres cannot be tackled at the communal level. A broader perspective is needed, especially when the concerned communes are just small satellites of a large city, as is often the case.

The promoter, *Coop Piemonte*, is part of the national group *Coop*, which is a leading company of large-scale retail distribution in Italy. The case shows that this promoter was interested in the profits originated by the commercial activity of the centre, and not in the economic windfall that would have been generated by the use of the land for a shopping centre instead of industry or agriculture. In fact, the project land was sold by the original owner to *Coop Piemonte* only after the reclassification of the property for the shopping centre use. Prior to the final choice of the site, the promoter was conducting a negotiation with three landowners simultaneously, presumably to exercise a stronger bargaining power in each case and in order to close the first deal that could be considered profitable in a fast-growing market where the firstcomer would certainly have gained a competitive advantage, just through being the first one to choose among the best locations.

Conclusions

This case warrants evaluation both on its merits and as an example of the continuing structural change in the retail market in Italy.

On its merits, the shopping centre *Le Fornaci* of Beinasco near Turin can be considered a success, because it was developed in a comparatively short time: six years, from July 1982, the date of the initial request by the promoter for a planning permission to open and operate a shopping centre, to June 1988, the date of completion of the construction works. Starting from the purchase of the land of the project site from the landowner by the developer in July 1985, the time is just three years. The centre of Beinasco has become a serious competitor for all the major shopping facilities of southern Turin. As of 1996, the promoter was considering an application for planning permission to expand the centre. This testifies to the correct choice of the location of the existing facility. With regard to the effects on the surrounding area, it must also be recognized that *Le Fornaci* fostered the spur of new developments in the fragmented outskirts of Turin such as the Commune of Beinasco. The flip side of the coin is that the success of this centre demonstrates that the plan of the *Comprensorio* was wrong in its classification of *Le Fornaci* as a simple neighbourhood shopping facility. Indeed, it seems that the importance of consistency in the urban planning action at the local level was underestimated, not so much within the same local body (sometimes it is already a good result), but across the major local government bodies, which have overlapping or interdependent responsibilities. In Italy an economic impact analysis of each new proposed shopping centre on the local retail sector is not yet mandatory. To be accepted, the private proposals need to fit in with the location criteria set at regional level: it is thus presumed that the approved shopping centres are economically compatible with the existing shopping network. If this is not the case, however, it is the private promoter and the concerned commune, not the institution *regione*, that have to bear the burden of the wrong location.

As part of the *convenzione*, the agreement signed with the Commune of Beinasco in order to obtain the planning permit, the promoter *Coop Piemonte* included in the project several landscape works (e.g. a road link between the parking area and the adjacent main road, public lighting of the whole area, restrooms and covered bus stops) that made the area more visible and attractive. One may argue that, in the end, the promoter was the primary beneficiary of the public works done as partial payment of the tax due for the issue of the planning permit for the construction of the shopping centre. This is a rather common event, as was underlined in other cases described in this book. Perhaps, it would be a good idea to revise the concept of *oneri di urbanizzazione*, so that the revenues collected in the form of public works would benefit not only the area around the project at hand but also other parts of the commune. This would allow a more equitable distribution of the benefits generated by the urban growth in certain locations of the commune.

Until the early 1990s (that is, before the new local administrations elected in 1993), the critical factor in the retail property business in Italy was to obtain an initial planning permit for a shopping centre of any size. Future lobbying with the local administrations, frequently supported by shaky political majorities, would have led to progressive increases of the authorized floorspace. In several cases, the promoter would have acted in the absence of either the planning permission to build the walls of the centre or the required licence to open and operate a full shopping centre, for in Italy there is a two-tier licensing process: one type of permit (granted by the commune) for the physical construction, and another type (granted by the region) for the retail activity. In both cases, in practice, there are not two single permits, but rather two streams of permits that usually facilitate, on one hand, step-by-step enlargement of the authorized floorspace and, on the other, the types of merchandise and services that can be sold by a given shopping centre.

In most cases the promoters did have the required authorizations, but often for smaller surfaces than those actually built and utilized. In these cases, the promoter used to run a calculated risk. The prescribed demolition of unlawfully built shopping structures was a remote venture, usually converted in a financial penalty in favour of the concerned commune. Also, the infringement of the strict rules that regulate the vending of goods, according to various merchandise classes, could in the end be converted into financial penalties. In both cases, the penalties would have been paid by a fraction of the added profits obtained through the non-yet-authorized shopping surfaces. Today, those old tricks are unlikely to succeed, for several reasons: there are new regional laws (e.g. the legislation approved by the region of Piedmont in 1995) which should prove more difficult to by-pass; there is an increased attention to the implementation phase of the existing laws; and the local administrators elected with the electoral law of 1993 have demonstrated an increased awareness in the field of urban planning and, in addition, have to justify their choices directly to the voters, as the role of the political parties has weakened.

It cannot be over-emphasized that the blind following of fashions is a risky business in property development as well as in urban planning. In practice, projects such as the one opened in Beinasco in 1988 were the bread and butter of the retail property market in North America some 25–30 years before. If this is an accurate comparison, one may wonder why, in the age of easy air travels and the Internet, the imitation activities in the property sector should be conducted with a time-lag of some decades. If an emulation activity may be accepted in a creative trial and error process, one should perhaps consider the best and most innovative examples, rather than the ordinary ones.

For example, while the shopping centre in Beinasco near Turin was being implemented, in San Diego (California) the opening of the Horton Plaza centre in 1986 marked the beginning of a new generation of shopping centres with strong entertainment components. The Horton Plaza centre has five floors above the ground, about 110000 m^2 of gross floorspace, 4 department stores and 136 specialized businesses (including a gym, a theatre and several restaurants). The idea of restructuring that area in the inner part of San Diego dates back to 1969, whereas the decision that proved to be the beginning of the real implementation process was taken in 1975 when a public–private partnership called Centre City Development Corporation (CCDC) was set up. The mission of this corporation was to revitalize downtown San Diego after the flight of residents and businesses that characterized most large cities in the USA during the 1960s and 1970s. The city administration played an active promotional role that could be defined nowadays as a successful urban marketing action. The development corporation was able to define a comprehensive plan for the renovation of the area around the proposed project, thus assuring that the shopping centre would have benefited from the type of urban environment, and negotiated in positive terms with Hahn Company, the promoter of the shopping centre.

Since its completion in 1986 (that is, two years before the opening of *Le Fornaci*), the San Diego centre has become a lively meeting place for youngsters, professionals and pensioners with time to kill. The centre is highly successful as a pure shopping facility, and its architecture is so unconventional and interesting to be praised in discerning architectural reviews. But the truly innovative fact is its location in downtown San Diego. Indeed, Horton Plaza has become a driving factor in the restoring of quality life in downtown San Diego, as is shown by the revival of the adjacent historic "Gas Lamp" neighbourhood and by the new Marina district.

When considering the retail market, Italy is usually regarded as trailing other major countries on the grounds that large shopping centres are not the prevailing form of retailing. That remark is based on the assumption that there would be a natural trend towards the restructuring of the retail sector in favour of few large shopping points. It is true that, since the early 1980s, the northern and central parts of the country have seen a remarkable growth in the number of shopping centres. These were initially accepted by the local administrations without major objections, as signs of modernity and because they brought new jobs. In a very early phase, the new centres were accepted with an innocent enthusiasm, as the one that accompanied the location of industrial plants in agricultural communes plagued by unemployment. The creation of new jobs was a persuading argument for the mayors of the small communes in the outskirts of large cities, where the majority of the new centres were built.

Recently, however, the local administrations are more and more cautious in granting planning permission for new centres, because there is a new generation of regional laws that strictly regulate the field. Indeed, what is changing is the general attitude towards the large shopping centres, for several reasons. The shopping centres are seen as traffic generators, and are indeed an added cause of pollution. It is now clear that there is a direct link between the opening of new large centres on the outskirts and the closing of traditional retail stores in the urban centres. The new centres tend to be built on the outskirts on green land, thus consuming a scarce resource and fostering a model of urban growth based on quantitative development, when public opinion is increasingly concerned with the quality of life.

The unnecessary consumption of green land stands in contrast with the general concept of sustainable urban development fostered by authoritative urban planners and also by many reports from the European Union (e.g. the *Green Paper on the urban environment*, and also *Europe 2000+*). The 5th Action Programme on the Environment (EC 1993, *Towards sustainability – a European Community programme of policy and action in relation to the environment and sustainable development*) has sanctioned once and for all the economic and social factors that may result as an outcome the urban sprawl in Europe. The out-of-town location of shopping centres was maintained to be in full contrast with the needs of European cities of the 1990s by the conclusions of research on the sustainable city completed by COREP – Polytechnic of Turin and CELP – London School of Economics (with funding from the European Union) at the end of 1994. That research analyzed the retail policy of two matched pairs of cities in Italy (Bologna and Florence) and the UK (Leicester and Edinburgh) and found that the out-of-town location of shopping centres is a culturally alien component in terms of the traditional urban policies of Italy and, to a lesser extent, of the UK. In addition, the development of shopping centres on agricultural land seems not to be in the medium to long-term interest of the promoters nor of local government. It would appear wiser, on one hand, to implement urban policies for the preservation of green areas and open spaces around the already-built urban cores, and on the other hand to facilitate brownfield projects, that is the recycling of old industrial areas for new developments, including shopping centres. Indeed, the presence of shopping centres of the latest generation (i.e. entertainment parks with different shopping facilities) could support the economic feasibility of lively mixed-use projects on old industrial land, certainly a better choice than monotonous office projects that are often unfeasible from a market viewpoint.

In addition to these arguments, it should be noted that in some regions of Italy (such as Emilia-Romagna, Lombardy and Piedmont), there are already signs of market saturation, as happens, for example, in many areas of England and France. Hence, there are serious chances that any newly approved shopping centre would go out of business in a few years. This would leave behind empty cement boxes built on previously green land: not exactly a desirable outcome for any wise urban planning policy.

Shopping centres attract citizens to shop inside cement boxes, where fake "Italian" streets and "Italian" squares are built for the attraction and rest of consumers. This is a model that can well fit the vast plains of the USA and Canada, where low-density settlements coupled with the low quality of the inner-city areas may give rise to a need for meeting points where people can meet each other in a safe environment with convenient access by car. It is a model far less convincing where outside the proposed "cement box" there are plenty of real Italian streets and Italian *piazze*, full of people strolling with pleasure or bustling for work, and not just consumers in a perennial quest for the lowest prices.

The latest generation of shopping centres in North America are closer to the concept of "entertainment parks", rather than to the standard layout of pure shopping facilities. Indeed, crowding is an effect sought by the promoter to generate the feeling of being in "a place" with a distinctive character, something that does not require elaborate explanation in terms of its location in the city, because it has risen to prominence in the mental maps of the active population of a given area. This is obviously very different from the "non-places" that have flourished in the panorama of urban America and, unfortunately, also in many other parts of the globe. But not in Italy, or at least not in the vast majority of small and medium-size cities (which are the backbone of the country) and in the historical parts of the largest communes. However, the peripheral settlements, built from the 1950s until the 1970s in the outskirts of the metropolitan centres of the country, can compete well with the worst modern architecture of this century.

In Italy, the standard model of out-of-town cement-box shopping centre, when replicated on a massive scale, can draw people away from the real streets and squares. Deserted streets are unpleasant and unsafe. Ultimately, a large number of shopping centres can be the cause of a drastic decline of entire urban districts. It has been maintained (Jacobs, 1961) that the net of small stores gives to a city the benefit of a social control that costs nothing but is very effective, both in preventing petty crimes on the streets and, above all, in creating the sense of a living community, which is a goal for all cities. Jacobs pleaded for urban districts to be rich in mixed primary uses, and this can be taken as a direct reference to the European city and, notably, to the model of the Italian city, with its streets full of warm life, colour, noise and diversity. It seems unlikely that the Italian local administrators will forget their own history and end up embracing a retail model that was conceived for the low-density and car-driven urban centres of North America. But perhaps there are no limits to the worst.

PART III
Conclusions

CHAPTER 7

Evaluation of the functioning of land and property markets in Italy

7.1 The Italian property sector in the Single European market

The European Single Market has been in place since 1993, but the Italian land and property market is far from being open to developers and investors from abroad. The property sector is in the middle of a structural transition from a time when national building firms and promoters had to face little or no competition, to a phase where competition with companies based abroad will become the rule, at least for business above a certain financial threshold. At present, however, the Italian property sector is largely heterogeneous, with a handful of companies active worldwide, whereas the vast majority usually do not go beyond a few Italian regions.

In the 1980s the Italian industrial system went through profound restructuring that resulted in a general increase in efficiency, and in a renewed appreciation of products "made in Italy" in world markets. Undoubtedly, the oil shocks of the 1970s threatened the very existence of entire sectors of Italian industry. We can clearly see now that the response was up to the challenge. But the property sector was shielded from most of the threats connected with the energy crisis and thus changed less or not at all. It is true to say that, since the first oil crisis of the early 1970s, the building industry has paid great and increasing attention to energy-saving materials and construction techniques. However, the construction phase is only a part of the whole development process, which in Italy has not significantly improved upon its traditional methods of doing business.

The Italian industry, even at medium enterprise level, was forced to introduce budget planning, strategic marketing concepts, "just in time" management of parts and supplies, and other advanced techniques. But in the property sector, feasibility studies, rates of return analysis, risk assessment - to name just a few techniques - are utilized only by some companies. Likewise, in the public sector, the strategic marketing approach to urban management is basically unknown, whereas elsewhere in Europe many local authorities are experimenting with it, and a few have already gained considerable experience in attracting foreign investors to implement local plans.

It seems that the majority of Italian cities and regions still do not see any new role for themselves other than the task of controlling development. As the turn of century approaches, the challenge for local government bodies is to "reinvent government" (Osborne & Gaebler 1993). Italian cities and regions should assume a new role of promoting local economic and social development through urban planning programmes, having three main guiding principles: strategy, equity and sustainability. It is the "3S" urban marketing concept, which stands for strategic, social and sustainable planning actions. According to the proponents of this urban marketing approach to city development (Ave & Corsico 1994), "3S" urban marketing should replace whenever possible the rigid top-down approach to planning, incapable of providing satisfactory answers in a world changing more rapidly than ever.

There are several issues that need to be addressed if the Italian property sector is to become more efficient and capable of withstanding competition in the context of the Single European Market.

7.2 Five issues which await action

First, the Italian fiscal system concerning property should be fundamentally reformed. It is widely felt that local authorities at the commune level should be given a greater degree of responsibility in taxing incomes from property. Obviously, such a reform might cause some new problems. For example, the local authorities might see it as an incentive to attract rich households and firms in order to increase their tax base. A phenomenon of this sort is fairly common in the USA, where the availability of funds for primary education is inextricably linked to the ability of local authorities to raise taxes from property. But this should not discourage the search for fairness and efficiency in the Italian system of property taxation.

The introduction in 1993 of *ICI* (the tax on real estate properties) was a first step in the right direction. Although this tax has been criticized by experts from all political parties, it seems unlikely that any future government will go back to the times when owning property assets was basically tax free. The question for the communes is how to use in a more flexible way (with regard to the tax rates and the issue of tax incentives) the fiscal tool to sustain the implementation of the city master plan.

Secondly, there is little connection between land-use planning and transport planning at local authority level. Transport networks can play a key role in opening up new areas for development or in sustaining market values of existing properties. But major networks (subways, highways, ringroads and the like) require heavy investment, well beyond the financial resources of local authorities, and which are in any case beyond their portfolio. Conflicts between local (commune), intermediate (region) and central (state) government – often ruled by different party coalitions – are fairly common in Italy, and result in "stop/go" financial policies with frequent stalemates.

At the city council level the uncertainty about state financial support for certain transport networks makes local urban planning a risky business, more an art than a science. In practice, many policy options concerning local urban planning are totally subordinate to the ordinary financial movements from the state to regions and communes, although in recent years financial disbursements for one-off events (the earthquake in the South, World Cup soccer in 12 cities, the Columbus fifth centennial celebration in Genoa, and so on) resulted in the construction of major transport networks. These ultimately encouraged private investment in the property market.

The successful planning of local authorities is often only a "pro-growth" planning activity aimed at property-owners and financiers. However, some city administrators have become aware that the time of pure and simple pro-growth planning policies is over for good. Physical expansion of the city is not always seen as "the right thing". Whether or not a commune should aspire towards growth is a question more and more frequently posed among city administrators, as it is now evident that physical growth not only has benefits but also sizable costs. Thus, the concept of "sustainable growth" is currently high on the agenda of urban planning researchers, as well as on the political programme of a small but growing number of city administrators.

Pro-growth planning should be seen as obsolete not only in Italy, but in all the advanced countries where development has already moved beyond a pure industrial phase. The progress of new technologies and of new production methods has greatly reduced the land needs of all sectors of industry. In addition, certain manufacturing operations have been physically moved in less advanced regions of the developed world.

Therefore, starting in the late 1970s, Western cities have created plenty of industrial wasteland. It must be underlined that these properties have surged to the interest of local administrators and real estate promoters for two basic reasons. First, the urban industrial areas represent for many local governments the only available areas where strategic services for the whole city can be located. Secondly, in the land market these properties are seen simply as urban land with often good locations and accessibility, and not merely as industrial land. In fact, there is much less interest in vacant industrial areas located outside the city. In the case of urban industrial land, the attraction is accessibility to the urban services and to the city itself. Hence, the property market aims to benefit from the potential windfalls that have accumulated on those areas as a consequence of recurrent public investments in urban infrastructures over time. In conclusion, it should not be forgotten that the source of "value" for the industrial wasteland, is given primarily by two factors: the ranking of each city compared to its competitor urban centres and the accessibility of each property with regard to the main transport links. As all these factors depend on public policies and investments, the local administrations must have a primary role in the decision on what to do with industrial wasteland and should also be assigned a sizable chunk of the market values generated by the re-use of those areas.

The re-use of vacant industrial and urban land is both the greatest opportunity and the greatest threat to Italian cities. What has to be decided is the use of the last available large plots in the hands of individual owners. There are no alternatives to re-use, as the expansion of urban centres on surrounding agricultural land does not appear to be a sustainable choice. The old industrial properties are ideal for the location of much-needed strategic urban infrastructure. The threat comes from the proposals for re-use that fall short on the grounds of social equity or market perspectives. There are two extremes to be avoided: first, to develop industrial wasteland with shopping centres and housing projects which would undermine any strategy for the development of the city; secondly, the use of industrial land exclusively as green areas, which would impose on future city administrations an excessive financial burden in maintenance costs. The re-use of industrial wasteland would normally be the best option for any Italian city.

The third issue to be emphasized is the fragmentation of the decision-making process in the Italian political system. This book has perhaps given a hint of the "legislative jungle" that flourishes in the field of urban planning in Italy. For a project to take off, it has to be approved by various levels of government and by several departments within each level.

After some widely publicized scandals connected to development projects in the 1980s, city administrators now have a more conservative attitude and give the green light to a project only when all but the most insignificant formal requirements are fulfilled. Italian cities are usually governed by coalitions of several parties, all having in various degrees a certain power of veto. Any major development project needs the agreement of the ruling coalition before being officially approved by the city council. Stalemates are common, because negotiations between political parties are never on a single issue – the project on the table – rather, they simultaneously encompass official urban policies and also the objectives of every single party in the urban field and other sectors.

The decision process in the urban planning field needs to be reported entirely in the local government assemblies, near the problems of firms and citizens, and kept far away from the private meeting rooms of political parties and enterprises. Conversely, it must be kept far away from sterile cultural debates where the real objective is not the definition of positive planning goals but the maintenance of narrow sectoral interests. Urban planning needs its sense of vision and its historic role of a key factor in the cultural and economic growth of society should be upheld. Urban planning needs to be directed again to the main problems of society, and not reduced to a branch of law studies, because to be productive any legislation in the urban market and planning fields must be just the last link in a chain of social changes on which there is a consensus among local communities.

These changes are the outcome of cultural positions and ideals on the final image that it is wanted for the city, and negotiations on more practical issues like the distribution of urban property values generated by the city master plan. But ultimately what really matters is whether the outcome of the development process is a city in which people and businesses want to live and work, or a city from which they wish to escape.

To make real progress, planners and decision-makers should abandon the tacit assumption that every problem, even a minor one, needs to be solved through a new law tailored to that specific issue. Indeed, there are already plenty of laws, and the real question is to find out how to implement existing legislation rather than to increase the number of laws. Planners and public administrators should find the way to think and act positively instead of wasting their best energies in fighting for unrealistic goals, such as the general expropriation of land or the curbing of planning gains through national laws, to the point that real estate activities would be totally discouraged or illegal practices would have to be silently accepted in order to let the real estate market function.

A recurrent request to local government by Italian building firms and promoters is to establish a "single desk" (*sportello unico*) where the necessary authorizations could be requested and obtained for all the projects in the pipeline. Action along these lines is a long way off. At present at the commune level, there is no special office in charge of examining proposals concerning large offices, tourism or light industrial projects. In theory, and sometimes also in practice, the representative of a foreign investor is expected to stand in the same queue as the private owner wishing to obtain a planning permit for a small project.

Local governments throughout Europe are charged with increasing responsibilities. At the same time, central governments demand that cities and regions become financially independent, and show equity and efficiency, not only in providing public services, but also in levying taxes. Under these circumstances, it is quite normal for local governments to move towards urban marketing.

However, my analysis of urban land and property markets in Italy has highlighted that there is a missing link between urban planning and tax revenue planning at the local level. Indeed, it seems that this link is also missing, or is rather weak, in other major countries. The planning and finance departments of large cities should cease to work separately. The urban master plan is to be seen also as a policy tool in the forecast of financial tax revenues generated at the local level within the period of validity of the urban plan. The outcome would depend on the type and quantity of property development foreseen in the land use map of a given city. Local governments would eventually end up by having a better idea of the amount of tax resources available for investment in local public services. Hence, the provision of public services could be planned more efficiently and this would improve the image of the location, generating a virtuous circle: a greater attraction for firms, institutions and households, and an expansion of the local property tax base, which would finance more local services, which would in turn act as new magnets for inward investments.

A fourth element is the considerable uncertainty about "time" in the development business. From the first three points it should be evident that in the Italian urban planning field there is an extremely low profile given to the decision-making process. The same is true for the land and urban markets, where sources of information are scarce and become basically unreliable when data on selling prices are sought, especially in the land market. Italian local government until recently has been in the midst of a legislative vacuum concerning expropriation rights. New laws in the 1990s have clarified the situation, but the problem is not entirely solved if local government cannot rely upon sufficient financial resources for land and property dealings. In the 1980s the legal vacuum led both private firms and public bodies to a "negotiated-approach" to urban planning. They saw it as a tool to get out of the development freeze caused by the absence of a clear legislation on land-use and development rights. But the negotiation procedures put into practice to by-pass the legislative vacuum varied from one region to another, and often from one city to the adjacent urban area. The result was even more confusion and a lower profile for urban planning.

In the property development business, time is obviously a crucial factor. All the promoters surveyed by the field research for this book conveyed the opinion that the ability to forecast the timespan for a development is more important than the absolute length of time required, say, to get building permission. A lengthy but predictable timespan is always preferable to an uncertain waiting period. In Italy, building concessions are not only lengthy to obtain, but are above all uncertain, as the approval procedure can jam at any stage of the decision-making process.

The fifth issue that awaits action is weak implementation capability in the planning field. As explained earlier, there are many factors that produce uncertainty in the implementation process of urban planning in Italy. Some of these factors have been tackled by recent legislative changes, starting with the new system (since 1993) of direct election to city and regional councils, which has provided an unprecedented stability in local governments and ultimately a better chance of actually implementing what the city master plan has just foreseen. So, there is hope.

However, the implementation capability has weakened since the 1960s in parallel with the weakening of the functions placed under the control of architects and planners and the increasing role accorded to lawyers. In fact, it can appear that the decision process in the planning field has shifted into the hands of greedy lawyers, whose objective interests are to perpetuate conflicts, not to reach agreements. Lawyers are typically more interested in stating principles than in checking whether or not the asserted rules can be implemented once they have been printed.

There is evidence from the innumerable laws that remain on paper, that those who propose ever more planning laws do not take the trouble of ensuring that the basic conditions required for implementation are there. After all, if the process fails because of the inadequacy of the laws (as often happens), this is a source of potential conflicts, creating work for lawyers. From this viewpoint, it is clear why the direct negotiation approach to urban planning has attracted a growing interest, when it is conceived as an open and fair search for the best balance between private profit and social equity.

The case studies have shown that property values are ultimately created by urban planning and investments in public infrastructures, especially transport infrastructures. Negotiations between local government and private developers were found to focus basically on three issues which constitute the primary rules of land and property markets:

- the destination of use of urban land
- the volume of construction authorized at each location through the system of planning permissions granted in accordance with a city master plan
- the type, quality and quantity of public investments directed to each area to support urban growth and ultimately property values.

The absence of efficient planning practices jeopardizes the stability and growth prospects of property values, and not only the well being of communities. This was found to be true of Italy, but it is likely to be a common feature in most other Western nations; thus the above-mentioned basic rules must be supported by all those interested in achieving best urban planning practices and fair property markets in their own country.

The case studies have also proved that accessibility is the single most important factor that affects property values, once both destination of use and construction volumes have been fixed by the urban plan. Although location is a natural feature, accessibility is an artificial condition influenced by the action of the different levels of government. Location does not change over time, but accessibility can vary greatly, as the result of public resources invested in transport infrastructures affecting either the location of development proposals, or adjacent locations.

The case studies of the aborted re-use of the industrial wasteland of Portello in Milan and the market failure of the restoration of the antique fortress of Baluardo in the port of Genoa, were clear examples of the dead-end to which the conflict between public and private interests can lead if there is no mutual trust between public administrations and private firms. Successful property development projects will depend more and more on true cooperation between all the actors involved, with a very pragmatic view from the public side and an open-minded attitude on the part of the private actors.

When solutions to urban planning issues are sought in law courts, it is probably a defeat for the whole city, because a huge quantity of resources are wasted (not least, time), while the city loses opportunities and confidence in its future. But to avoid fruitless confrontations, which are bound to end with no real winner, there is the need for clear goals in the urban planning field and for an authoritative guidance from the local government in the development process. In this framework, the city master plan is bound to play a crucial role, provided it is conceived and implemented not merely as a defensive weapon against any private development project, but as a proactive tool to foster the social and economic growth of all components of a local community.

7.3 Conclusions

Urban planning arose as a modern discipline to regulate the construction of the city in the first industrial revolution, in the nineteenth century. Some have thought that, approaching the end of the twentieth century, the role of urban planning was over for good. I think this is wrong. In perspective, urban planning will be needed more and more, as cities and urban systems are bound to become ever more complex configurations of networks and flows and relationships between residents, firms and institutions. The poor results of the rejection of planning during the Reagan and Thatcher years should suggest that planning is always the cornerstone of the creation of property value, as only urban planning can introduce scarcity in the urban land and property markets.

In Italy, there is a strong need for a true reform of the general planning legislation. At present, various proposals have been put forward. It seems likely that the new government elected in Italy on the 21 April 1996 (a centre–left alliance called *Ulivo* and led by Romano Prodi) will address this issue, with a good chance of achieving an overall reform. In fact, in the winning coalition the *PDS* (the democratic party of the left, formerly the Italian Communist Party), plays a leading role because it received most votes (over 21% of the total). For example, the majority of the new ministerial positions have been awarded to the *PDS*. In the new government, the important Ministry of Public Works has been assigned to Antonio Di Pietro, the former judge who, between early 1992 and the end of 1994, had become a national symbol of the fight against corruption in the public works sector.

In the past, the main proposals to reform physical planning legislation have been put forward by individual experts and cultural institutions (such as *INU*, *Istituto Nazionale di Urbanistica*, the national institute of city planning) close to the political parties that are now in power. Thus, there are chances that the cornerstone of the Italian physical planning system, the *piano regolatore* or *PRG* (the city master plan) will be reformed along the lines of these proposals. In brief, the *PRG* is likely to be replaced by a physical structure plan that would concern solely the main urban infrastructures and the location of the key urban functions, while the day-to-day management of the development process would be dealt with by more flexible planning tools.

In Italy, it is likely that the urban land and property markets at the turn of the century and in the following two or three decades will be dominated by the issue of re-use of old industrial areas. This will probably be accompanied by the issue of re-use for vacant out-of-town shopping centres, together with the current problem of re-use of office properties. The key point is that Italian cities, like most western European cities, abound with empty industrial areas, whose utilization would flood the property markets with the possible real estate products. Urban planning is the only way to create conditions of scarcity in the supply of properties; thus, it is only through planning that property market values can be effectively created and sustained over time. The recognition of this fact could well become the basis for common ground between local administrations and private firms.

Ironically enough, the cumbersome Italian planning system has at least one unexpected positive feature. Development proposals remain on the discussion table for long periods, which gives everyone a chance to have a say in the decision. Large development projects go through a sort of "decanting process" in which the discretional power of local authorities can be used to give more time and space to the interests of the local community or, conversely, to stop a project altogether, which is sometimes the "second best" option.

The lack of formal organization typical of the Italian property market can have the effect of acting as a valid barrier to external penetration. This might be seen as a positive feature by Italian companies who fear competition. It is obviously a very short-sighted attitude; nevertheless, it is there and we should be aware of it. But the other side of the coin is that the same low profile might discourage foreign investors from choosing Italy, no matter how effective the urban marketing system might be. Italian local authorities are beginning to realize the necessity of positive action to address these problems, but a formidable task lies ahead.

It seems that we all ought to focus our attention again primarily on the problems of real cities instead of the intricacies of the urban plans, on actual needs of firms and citizens, instead of formalized and rigid standards for public services. Urban analysts and planners should stop seeing the city from the viewpoint of urban planning and its internal requirements, and take the risk of reversing their attitude and continuously reform urban planning according to the viewpoints of market demand of firms and institutions and social needs of citizens. In short, urban planners and decision-makers must rediscover a true interest in actual cities, with all their beauty, chaos, noise, diversity, and all the facets of real life.

Appendix

The main legislative acts of the central government, affecting urban land and property markets in Italy. Numbers in **bold** refer to the legislation number.

2359, 25 June 1865 (expropriation)
First law of united Italy to establish the right to expropriation for public use and to determine the compensation amount for the taken properties. Still in effect.

2892, 15 January 1885 (expropriation)
Introduced the distinction between "developable land" and "agricultural land", which has been accepted by all subsequent legislative acts. It introduced more restrictive criteria to determine the compensation amount for expropriated properties. The compensation was set equal to the average of the actual value of the property plus the sum of the rents in the previous ten years. It became known as *Legge di Napoli*, Naples Act, because it was enacted also to allow the renovation of the city of Naples after an epidemic.

1089, 1 June 1939 (listed buildings)
The constructions having recognized historic values are entered in special lists of "buildings to be preserved". Any kind of alteration and real estate activity on these buildings must obtain, on top of the regular approvals, the *nulla osta*, the endorsement of the *Sovrintendenza*, the state agency for architectural and cultural beauties. In case of sale to a private party of a listed property, the state has the right to declare void the transaction and buy the property at the same price, within 2 years from the date of the sale.

1497, 29 June 1939 (natural beauties)
Law to preserve natural beauties and real estate properties having special features. It required a *Piano territoriale paesistico*, a landscape plan.

1150, 17 August 1942 (main physical planning act)
Planning is implemented through *Piani regolatori*, physical master plans, at commune level and between communes. Cities within a special list are obliged to have a master plan, while for the rest of the communes it is an option. All communes must have a *Regolamento edilizio*, a building code. When a commune doe not have a master plan, the building code has to include a *Programma di fabbricazione*, a programme of the real estate activities. Master plans at commune level are implemented through *Piani particolareggiati*, detailed plans. To implement the master plan, a commune can expropriate private properties, which can also be taken when in contrast with the provisions of the master plan. All real estate activities are made subject to a licence issued by the mayor of the concerned commune.

1902, 3 November 1952 (safeguard)
A commune can suspend the exam of the building licences which are deemed in contrast with a master plan which is being prepared, in the period in between the decision of the commune to approve the master plan and its formal approval by a superior government body (the Ministry of Public Works until 1970 and the regions thereafter).

167, 18 April 1962 (main law for public housing)
Obligation for the communes with population above 50000 people, and other communes with special characteristics, to have a plan for the construction of public housing. The land required for these plans is to be expropriated.

1187, 19 November 1968 (safeguard)
The ban to privately develop a land parcel subjected to public use as indicated by the master plan, is given a time limit of 5 years (previously there was no time limit). If, within said limit, the *Piani particolareggiati*, the detailed plans, or the *Piani di lottizzazione*, the plans to subdivide into lots a property, are not approved, the ban loses any legal validity.

426, 11 June 1971 (main law for the retail network)
The communes having certain characteristics are obliged to draw a plan for the development and rationalization of the retail network, both mobile (open-air markets) and non-movable stores. The plan indicates for each commercial branch a threshold for the desirable retail space in the commune, and grants (or denies) the licences accordingly.

865, 22 October 1971 (housing and expropriation law)
Updates the contents and procedures of previous laws concerning the expropriation process for private land taken for public uses. On the land parcels expropriated for public housing, the commune can grant the free-hold to public agencies for a period between 60 and 99 years, to build public housing as well as public services. It introduces the *Piani insediamenti produttivi*, plans for artisans and light industrial buildings. The communes without a plan for public housing can locate projects for public housing in the areas zoned as regular "housing" by the master plan.

10, 28 January 1977 (main law on land to be developed)
Any activity which brings to a urban change or to a real estate development is subjected to the payment of *oneri di urbanizzazione*, a "development tax" related to the costs to provide urban public services and to the cost of the development project. The tax has two components, one for primary public services (sewage, roads, etc.) and one to cover secondary public services (primary schools, green areas and sports facilities). The planning permit can be granted only after the payment of the "development tax". Updates to the rules to determine the compensation amount for the expropriated land. It introduces for communes of any size the *Programma pluriennale di attuazione* (PPA), the multi-year implementation programme which calls for an implementation in phases of the master plan.

457, 5 August 1978 (urban restoration)
All communes having certain features are required to indicate the urban zones to be restored through *Piani di recupero*, restoration plans, consistent with the city master plans. Private owners can propose restoration plans to the city administration. Fiscal incentives are foreseen to facilitate the restoration activities. It launched the *Piano decennale*, a ten-year plan to sustain the construction of public and private housing having certain characteristics.

392, 27 July 1978 (main fair rent act)
It put an end to the rent freeze which had been in effect, through many renewals, since the end of the Second World War. It fixed a ceiling to the rent charges considered legal, which could not exceed 3.85% of the conventional rental value of a property, as estimated according a long list of parameters (cadastral class, size, location, maintenance etc.). The fair rent so established could be increased, at request of the owner, every year at the rate of 75% of the official inflation index as published by *ISTAT*, the official statistics institution.

94, 25 March 1982 (private housing)
The communes with population below 10,000 people are made exempt from the obligation to have a *Programma pluriennale di attuazione* (*PPA*), the multi-year implementation programme, unless required by the region. Financial grants are foreseen for the construction or renovation of private housing having certain features. The beneficiaries must have certain social characteristics and be within given income ceilings. Housing built or renovated with such incentives cannot be sold for at least 7 years (more frequently 10 years).

47, 28 February 1985 (main penalty remission)
First law to forfeit penalties for those who built constructions not in line with city master plans or with some planning acts, upon payment of a fine. The amount of the fine was set greater for recent buildings and smaller for older buildings. Only the construction completed before the 1 October 1983 could qualify for the penalty remission. The issue of the penalty remission was to be refused by the mayor when the request concerned constructions built upon public land, or alterations to natural beauties as defined by law, or buildings to be preserved as indicated by the special lists prescribed by law. In these cases the illegal constructions had to be demolished. It introduced certain simplified procedures to obtain planning permissions for works concerning the internal parts of buildings.

52, 17 February 1985 (office of real estate deeds)
Introduction of office automation procedures in all the Italian *Conservatorie dei registri immobiliari*, the offices of the real estate deeds, responsible for storing the data of property transactions.

431, 8 August 1985 (environment protection)
All land areas within 300 metres from the sea and from the lakes are preserved. No construction can be allowed: within 150 metres from the rivers; above 1,600

metres in the Alps and 1,200 in the Appennines and in the islands; in the archeological zones, natural parks, in the wood areas (even when burned out by fires).

122, 24 March 1989 (parking buildings)
The 15 largest Italian cities, plus a number of smaller cities (with at least 50,000 people or touristic centres) indicated by the regions, had to prepare plans for the construction of underground parking structures. The projects could be implemented by private firms in public areas given in lease-hold for a number of years (99 years maximum) or by private owners of buildings with certain features. The construction of parking structures had to be carried out in three-year plans, with a correspondent financial aid from the State. The construction of parking facilities was made exempt from the payment of *oneri di urbanizzazione*, the development tax.

142, 8 June 1990 (main law of local government powers)
A number of innovative principles for the powers and responsibilities of local governments. A number of cities and related urban centres of ordinary regions (Turin, Milan, Venice, Genoa, Bologna, Florence, Rome, Bari, Naples) are defined "metropolitan cities". The provinces are given the duty to prepare a *Piano territoriale di coordinamento*, a plan to coordinate the planning action at provincial level.

240, 4 August 1990 (inter-modal platforms)
Established the five-year plan to design and implement in each region inter-modal platforms for the freight service.

179, 17 February 1992 (public housing and area planning)
New norms to facilitate the implementation of restoration plans, through the bypassing of the opposition to the plan made by owners of small properties included in a restoration plan. Introduction of the *Programmi integrati di intervento*, the integrated action plans, promoted by the communes or by private owners to restore urban zones, property clusters or environment areas.

285, 30 April 1992 (road code)
The new road code introduced a complementary planning tool called *Piano urbano del traffico*, the city plan for traffic control, to be prepared by the communes with population above 30000 people and also by the smaller ones with large touristic flows, within April 1993. The communes have to approve *Zone a traffico limitato*, restricted traffic areas, and *Aree pedonali urbane*, urban pedestrian zones, and were given the power to establish paying parking areas within the restricted traffic areas and in the zones adjacent to monuments, and of special interest.

359, 8 August 1992 (expropriation)
Update of previous laws, introducing the possibility for the owner subject of an expropriation act to obtain a 40% premium on the compensation amount in case of voluntarily hand over of the property, thus with a reduction of time required for the expropriation process.

Decree **649**, 25 November 1994 (second penalty remission)
The second penalty remission concerned the real estate projects which could have qualified for the first penalty remission of 1985 (see above) but for which no application was presented on time. It included also the projects built after the 1 October 1983, with the usual exception of the buildings built on public land, in protected areas (parks, river banks, etc.) or the properties having a recognized historic value.

216, 23 June 1995 (procurement rules for public works)
Update of **109** of 1994 (known as *Merloni* Law), with which new and stricter rules were issued concerning the procurement procedures of public bodies for the purchase of goods and services, after the abuses of the 1980s and early 1990s. The new law enabled to relaunch investments in the public works sector, which had fallen to an all time low in the period 1992–94.

Bibliography

Alberti, M., G. Solera, V. Tsetsi 1994. *La città sostenibile*. Milan: Franco Angeli.

Allione, M. 1976. *La pianificazione in Italia*. Venice: Marsilio.

ANCE 1993. *Nuove regole per la crescita urbana*. Rome: Edilstampa.

Ave, G. 1981. *Il problema della casa oggi in Italia*. Turin: Celid.

— 1992. La città sostenibile tra immigrazione e mercato immobiliare. In: *La città sostenibile*, E. Salzano (ed.), 35–48. Rome: Edizioni delle Autonomie.

— 1993. Quale urbanistica per la città impresa. *Paesaggio urbano* **3-4**, 25–31.

— 1995. Le città italiane da centri di costo a centri di profitto per fini sociali. *Diritto ed economia* **1**, 275–87.

Ave, G. & F. Corsico (eds) 1994. *Urban marketing in Europe*. Turin: Edizioni Torino Incontra.

Ave, G. & F. Prizzon 1977. Le immobiliari quotate in borsa. In *Il secondo ciclo edilizio*, A. Barp (ed.), 85–134. Milan: Franco Angeli.

Balducci, A. 1991. *Disegnare il futuro*. Bologna: Il Mulino.

Benevolo, L. 1979. *Urbanistica e crisi economica*. Bari: Laterza.

Bortolotti, L. 1978. *Storia della politica edilizia in Italia*. Rome: Editori Riuniti.

Calarco, F. 1995. *Espropriazione della proprietà privata per opere di pubblica utilità*. Milan: Pirola.

Campos Venuti, G., & F. Oliva (eds) 1993. *Cinquant'anni di urbanistica in Italia 1942–1992*. Bari: Laterza.

Castellano, A. & M. Vitta (eds) 1988. *Il nuovo portello a Milano – the new portello in Milan*. Milan: Arcaedizioni.

Ceccarelli, P. 1978. *La crisi del governo urbano*. Venice: Marsilio.

CENSIS 1984. *Lo shock edilizio*. Milan: Franco Angeli.

— 1989. *Torino metropoli internazionale*. Turin: ISEDI.

— 1990. *Italy today*. Rome:CENSIS.

— 1995. *29° rapporto sulla situazione sociale del paese*. Rome: CENSIS.

Cervellati, P. L. 1991. *La città bella*. Bologna: Il Mulino.

Colombo, G., F. Pagano, M. Rossetti (eds) 1995. *Manuale di urbanistica*. Milan: Pirola.

Coopers & Lybrand 1996. *Global real estate tax summary*. Amsterdam: Coopers & Lybrand.

Coppo, M. & M. Cremaschi (eds) 1994. *Strutture territoriali e questione abitativa*. Milan: Franco Angeli.

Corsico, F. & A. Fubini (eds) 1994. *Aree metropolitane in Europa*. Milan: Franco Angeli.

CRESME 1991. *La costruzione della città europea negli anni '80*. **1** & **2**. Rome: Credito Fondiario.

Crosta, P. L. & S. Graziosi (eds) 1977. *Chi decide la città*. Milan: CLUP.

Dandri, G. 1977. *Il deficit abitativo in Italia*. Milan: Giuffrè.

Dematteis, G. (ed.) 1992. *Il fenomeno urbano in Itlaia: interpretazioni, porspettive, politiche*. Milan: Franco Angeli.

Fondazione Giovanni Agnelli, Associazione per Tecnocity (eds) 1995. Catalogo dei progetti per Torino - 1995.*Contributi di ricerca*, **June**.

Gabetti 1996. Milan Metropolitan Area. In *ULI - the Urban Land Institute, ULI market profiles 1996: Europe*, *ULI*. 81–5. Washington, D.C.: ULI.

Gabrielli, B. 1993. *Il recupero della città esistente*. Milan: Etas.

Gambaro, A. 1977. *Proprietà privata e disciplina urbanistica*. Bologna: Zanichelli.

Gasparini, C. 1994. *L'attualità dell'urbanistica*. Milan: Etas.

Ginsborg, P. (ed.) 1994. *Stato dell'italia*. Milan: Il Saggiatore.

Giustiniani, C. 1981. *La casa promessa*. Turin: Einaudi.

Healey, P. (ed.) 1994. *Trends in development plan-making in European planning systems*. Working papers 42. Centre for Research in European Urban Environments, University of Newcastle.

Indovina, F. (ed.) 1972. *Lo spreco edilizio*. Padova: Marsilio.

— 1977. *Dal blocco dei fitti all'equo canone*. Venice: Marsilio.

— 1991. *La città di fine millennio*. Milan: Franco Angeli.

— 1993. *La città occasionale*. Milan: Franco Angeli.

INU-Istituto Nazionale di Urbanistica 1995. *Politiche urbane*. Rome: INU.

Jacobs, J. 1993. *The death and life of great american cities*. New York: The Modern Library.

Le Pera, L. 1995. Commento sistemico alle norme per la determinazione dell'indennintà di esproprio per p.u. delle aree edificabili. Art. 5 bis, legge 8 agosto 1992, n. 359. *Rivista del consulente tecnico* **1**, 57–78.

Mantini, P. & F. Oliva (eds) 1996. *La riforma urbanistica in Italia*. Milan: Pirola.

Mazza, L. 1986. Nuova centralità e nuove ideologie urbane. In *Verso una nuova centralità delle aree urbane nello sviluppo dell'occupazione*. G. Garofoli & I. Magnani (eds), 17–36. Milan: Franco Angeli.

Mazza, L. (ed.) 1988. *Le città del mondo e il futuro delle metropoli*. Partecipazioni Internazionali, XVII Triennale di Milano. Milan: Electa.

MEDEDIL & Comune di Napoli 1989. *Centro direzionale di Napoli*. Naples: MEDEDIL.

Milano Centrale 1996. *Progetto Bicocca CD-ROM*. Milan: Milano Centrale-Gruppe.

Muccini, P. 1994. Staying put in Milan. In *Euro-city surveys 1994*, 45–9. London: Europroperty.

Nomisma 1995, *Osservatorio sul Mercato Immobiliare, Rapporto 2*. Bologna: Nomisma

Osborne, D. & T. Gaebler 1993. *Reinventing government*. New York: Plume.

Pacini, M., K. R. Kunzmann, J. N. Marshall (eds) 1993. *La capitale reticolare*.

Turin: Fondazione Giovanni Agnelli.
Padovani, L. (ed.) 1988. *Politica o non politica della casa?* Milan: Franco Angeli.
Preite, M. 1979. *Edilizia in Italia*, Florence: Vallecchi.
Royal Town Planning Institute 1991. *The planning system of Italy*. Audio cassette. London: The Royal Town Planning Institute.
Secchi, B. 1974. *Squilibri regionali e sviluppo economico*. Padova: Marsilio.
— 1984. *Il racconto urbanistico*. Turin: Einaudi.
Secchi, B., C. Baianchetti, F. Infussi & U. Ischia (eds) 1987. *Progetto bicocca*. Milan: Edizioni Electa. (English text version).
Travaglia, S. 1994. *Come funziona l'italia*. Milan: Sperling & Kupfer Editori.
Vitali, O. 1990. *Mutamenti delle aree urbane*. Milan: Franco Angeli.

Index